Advanced Materials and Structures
for Extreme Operating Conditions

Jacek J. Skrzypek · Artur W. Ganczarski
Franco Rustichelli · Halina Egner

Advanced Materials and Structures for Extreme Operating Conditions

 Springer

Prof. Dr. Jacek J. Skrzypek
Assoc. Prof. Dr. Artur W. Ganczarski
Dr. Halina Egner

Politechnika Krakowska
Inst. of Applied Mechanics
and Machine Design
Al. Jana Pawla II 37
31-864 Krakow
Poland
Jacek.Skrzypek@pk.edu.pl
Artur.Ganczarski@pk.edu.pl
Halina.Egner@pk.edu.pl

Prof. Dr. Franco Rustichelli

Universtá Politecnica della Marche
Dipto. Scienze Applicate ai Sistemi Complessi
(SASC)
Via Brecce Bianche
60131 Ancona
Italy
f.rustichelli@univpm.it

ISBN 978-3-540-74299-9   e-ISBN 978-3-540-74300-2

DOI 10.1007/978-3-540-74300-2

Library of Congress Control Number: 2007939399

Typesetting: supplied by the authors
Production: LE-TEX Jelonek, Schmidt & Vöckler GbR, Leipzig, Germany
Cover design: eStudioCalamar S.L., F. Steinen-Broo, Girona, Spain

Printed on acid-free paper

9 8 7 6 5 4 3 2 1

springer.com

Introduction

Increasing industrial demands for high temperature applications, high temperature gradients, high heat cycle resistance, high wear resistance, impact resistance, etc., require application of new materials. Conventional metallic materials, such as steels, nickel- and aluminium-based alloys, etc. cannot resist such extreme operating conditions. They have to be replaced by new metal/matrix or ceramic/matrix composite materials, MMC or CMC, such as titanium/zirconia, titanium/alumina, nickel/zirconia, nickel/alumina, steel/chromium nitride MMCs, or titanium carbide/silicon carbide, alumina/zirconia CMCs, etc. For instance, zirconia-based materials are known for good thermal insulation, but are bad for oxygen insulation, alumina-based materials are good for oxygen insulation and relatively good for thermal insulation, chromium nitride materials are used for good wear resistance due to their high hardness and not so bad thermal insulation, to mention only some. In general, aluminium matrix composites have a useable temperature upward of 300°C, whereas titanium matrix composites can be safely used up to 800°C. For higher temperature applications, up to 2000°C or even more, ceramic matrix composites have to be used. In spite of good thermal, wear and anti-oxygen properties of ceramics, their mechanical properties are rather poor, mainly due to high brittleness and low toughness resistance.

The first methods to improve thermal, wear, oxygen, etc. resistance of metallic structures, without suppressing their strength and toughness, consisted in depositing thin layers of appropriate ceramics on the metallic substrate, to produce thermal, wear or oxygen barrier coatings T/W/O BCs. Thin layer deposition techniques, such as plasma spray deposition and electron beam physical vapour deposition, usually cause high compressive residual stresses in thin coating, whereas negligibly low tensile residual stresses occur in the metallic substrate. The magnitude of compressive residual stress in the coating depends on the deposition technique, temperature, pressure and other conditions. High compressive stress in the ceramic coatings, not only at the initial state but also during the whole loading programme, is desired for poor ceramic resistance to tensile stress damage and failure. For instance, in the case of repeated heat pulses, the phenomenon of temperature ratchetting may be observed, which may result in a drop in compressive stress magnitude in the coating, or even produce an inadmissible tensile stress in the ceramic. Additionally, the severe residual and working stress discontinuity at the coating–substrate interface usually causes damage to the coating or metallic substrate, or failure due to delamination of the ceramic film from the substrate.

As a remedy to the aforementioned disadvantages observed in thin-layered structures, a concept of functionally graded materials (FGMs) may be proposed, where a smooth spatial gradation of the material properties from that of ceramic to metal, is achieved. In this way a sharp properties mismatch is eliminated, such that stress and strain discontinuities can be mitigated in the structure and both plasticity and damage limitation is met.

Proper modelling of the composite materials, identification of the developed constitutive models for selected nonhomogeneous materials, application of the models to complex thermomechanical loading programmes and initial/boundary value problems, finite differences FD and finite element FE implementation of the models to existing computer codes, development and implementation of advanced measurements techniques for stress, damage and microstructure changes, are the main goals of this book.

Constitutive models have to be capable of capturing combined physical phenomena, such as thermo-elasticity, plasticity, damage, low cycle fatigue, thermal ratchetting on heating cycles, thermal fatigue, damage anisotropy, crack opening/closure, failure due to thermal stability loss, etc.

Numerical methods have to be adjusted to the coupled steady state or transient thermomechanical analysis, including residual stress due to the fabrication phase, material nonhomogeneity, temperature dependence of material data, local or nonlocal approaches, failure prediction by damage field analysis and the substitutive crack concept, instability due to hot-spots analysis, etc. Additionally, when the classical finite element method based on the homogeneous elements is used for FGMs, the material properties vary from element to element, in a piecewise continues fashion, since all integration points inside the element have a common property value. This approach may be insufficient in the case of severe properties mismatch at the layer–substrate interface or if properties change through a functionally graded structure. Hence, the problem of the graded final element to discretize property is also addressed.

Eventually, enhanced experimental methods, essentially based on the scattering of neutrons and X-rays, are developed and implemented for residual stress, damage and microstructure change analyses in specific functionally graded and composite structures. The techniques have to be capable of detecting properties not only at the surface, but also at a depth of some micrometres in the substrate material.

This book contains the following:

- survey of advanced material properties for high-temperature applications (Chapter 1);
- basis of thermodynamic modelling of advanced materials (Chapter 2);
- development of selected constitutive models of thermo-elasto-plastic-damage materials (Chapter 3);
- implementation of models for specific composite and functionally graded applications (Chapter 4);
- methodology and implementation of experimental microstructure analysis to real composite and FGMs application (Chapter 5);
- guide to European sources of neutron and X-ray synchrotron radiation facilities (Appendix).

It was possible to reach this objective when specialists in the fields of constitutive modelling of advanced materials, numerical methods and computer simulations, structural analysis by FDM and FEM, experimental microstructure and damage analysis and residual stress determination by developed neutron and X-ray diffraction techniques combined with three–dimensional imaging, co-operated in the Network of Excellence: *Knowledge Based Multicomponent Materials for Durable and Safe Performance* KMMNoE. Chapters 1 through 4 have been written by J. Skrzypek, A. Ganczarski and H. Egner (former name H. Kuna-Ciskał), all from the Cracow University of Technology, Institute of Applied Mechanics, CUT, Kraków, Poland. F. Rustichelli from Università Politecnica delle Marche, Sezione di Scienze Fisiche, UNIVPM, Ancona, Italy wrote Chapter 5 and the Appendix.

It is necessary to mention the long list of co-workers and graduate or PhD students, whose works, papers and master theses were the basis for numerous examples, computer simulations, experiments, etc., reported throughout the book. Among them, within the CUT group, we appreciate the important impact in the field of constitutive models development and implementation, as well as computer simulations, by the co-workers: Dr J. Bielski, (CUT), Dr W. Egner (CUT), Dr P. Foryś (CUT) and L. Barwacz. Important contribution to the book is also due to the works of PhD students and graduate students, obtained when preparing their master theses at CUT and writing numerous joint papers and conference presentations, namely: S. Hernik, M. Cegielski, M. Oleksy, M. Kula and M. Juchno. Their contributions are additionally related to the KMMNoE tasks, in particular: WPR2 *Metal Ceramic Composites*, NRT2-2 *New Particulate Al and Ti Matrix Composites Produced by Self Propagating High Temperature Synthesis* (in co-operation with INASMET-TECNALIA, Donostia-San Sebastián, Spain), and WPR3 *Functionally Graded Materials*, NRT3-5 *Stress Related Phenomena* (in co-operation with Werkstoff-Kompentenzzentrum-Leoben Forschungsgesellschaft m.b.H., Leoben, Austria). On the site of UNIVPM the following researchers, who brought their contribution when writing the book, are appreciated: Prof. Dr G. Albertini, Dr V. Calbucci, Dr F. Fiori, Dr E. Girardin, Dr A. Giuliani, Dr V. Komlev, Dr A. Manescu, Dr F. Spinozzi. Finally, the

VIII

enormous effort made by Prof. A. Ganczarski, who created the whole book-file in LATEX with the assistance of D. Skrzypek, who prepared a number of chapter-files and figures, is gratefully acknowledged.

Jacek Skrzypek, Ancona, August 2007

Acknowledgements

This monograph was supported in part by the EU Network of Excellence Project KMMNoE, Knowledge-based Multicomponent Materials for Durable and Safe Performance, No NMP3–CT–2004–502243.

The grant PB 807/T07/2003/25 from the Ministry of Science–State Committee for Scientific Research, Poland is gratefully acknowledged.

Contents

Notation

Operators

x – scalar

$\mathbf{x}$ – vector or tensor

$[\mathbf{X}]$ – matrix

$[\mathbf{X}]^{\mathrm{T}}$ – matrix transposed

$[\mathbf{X}]^{-1}$ – inverse matrix

$\det[\mathbf{X}]$ – determinant

$|x|$ – absolute value

$\langle x \rangle$ – Macauley bracket $= \begin{cases} x \text{ if } x \geq 0 \\ 0 \text{ if } x < 0 \end{cases}$

$\langle\langle \mathbf{x} \rangle\rangle$ – normalized Macauley bracket $= \begin{cases} 1 \text{ if } x \geq 0 \\ 0 \text{ if } x < 0 \end{cases}$

$\dfrac{\mathrm{d}}{\mathrm{d}x}$ – derivative

$\dot{x}$ – time derivative $= \dfrac{\mathrm{d}x}{\mathrm{d}t}$

$\dfrac{\partial}{\partial x}$ – partial derivative

∇ – nabla $= \dfrac{\partial}{\partial x} + \dfrac{\partial}{\partial y} + \dfrac{\partial}{\partial z}$

∇^2 – Laplace operator

$\overline{\nabla}^2_1$ – axisymmetric Laplace operator $= \dfrac{1}{r}\dfrac{\mathrm{d}}{\mathrm{d}r}\left(r\dfrac{\mathrm{d}}{\mathrm{d}r}\right) - \dfrac{1}{r^2}$

$\overline{\nabla}^2_k$ – generalized axisymmetric Laplace operator

$\quad = \dfrac{1}{r}\dfrac{\mathrm{d}}{\mathrm{d}r}\left(r\dfrac{\mathrm{d}}{\mathrm{d}r}\right) - \dfrac{k^2}{r^2}$

div   – divergence

grad – gradient

$\mathbf{x} \cdot \mathbf{y}$ – single contracted product of tensors

$\mathbf{x} : \mathbf{y}$ – double contracted product of tensors

$\mathrm{Tr}(\mathbf{x})$ – first invariant of tensor

$J_2(\mathbf{x})$ – second invariant of tensor

H – Heaviside function

$\widehat{H}$ – Hamilton's operator

$\mathbf{n}_i \otimes \mathbf{n}_j$ – tensor product
$\delta(x)$ – Dirac's delta function
$\ln x$ – Napier's logarithm
$\exp x$ – exponent function
$\arctan x$ – arcus tangens function
$\mathrm{Re} x$ – real part of complex variable

List of Latin and Greek Letters Used as Symbols

$$a \ c \ e \ g \ i \ k \ m \ p \ s \ u \ w \ y \ \alpha \ \varepsilon \ \vartheta \ \nu \ \varrho \ \phi \ \omega \ \Lambda$$
$$A \ C \ E \ G \ I \ K \ M \ P \ S \ U \ W \ Y \ \beta \ \zeta \ \kappa \ \xi \ \sigma \ \varphi \ \Gamma \ \Phi$$
$$b \ d \ f \ h \ j \ l \ n \ r \ t \ v \ x \ z \ \gamma \ \eta \ \lambda \ \pi \ \tau \ \chi \ \Delta \ \Psi$$
$$B \ D \ F \ H \ J \ L \ N \ R \ T \ V \ X \ Z \ \delta \ \theta \ \mu \ \rho \ \nu \ \psi \ \Theta \ \Omega$$

Symbols

a – material constant
a_x – acceleration along x direction
A – area
A – scattering length
A_j – vector of conjugate thermal force
$\dot{A}_{\mathrm{eq}}$ – rate of equivalent crack area
b – plastic material constant in Hayakawa–Murakami equation
b_X – scattering length of electron
B – thermodynamic force conjugate of β
$\overline{B}$ – nonlocal force conjugate
B_0 – material constant in damage dissipation potential
B_j – vector of conjugate thermal force
$B_{ijkl}^{(\varepsilon)}$ – fourth-rank unilateral effect strain transformation tensor
$B_{ijkl}^{(\sigma)}$ – fourth-rank unilateral effect stress transformation tensor
$\mathcal{B}$ – membrane stiffness
$\mathbf{B}$ – magnetic field vector
$\mathbf{B}^{(e)}$ – element transformation matrix
$\mathbf{B}^{\mathrm{i}}$ – Eshelby's tensor
c – material constant
c – light speed in vacuum
c – particle concentration
c – specific heat
c_{c} – specific heat of ceramic
c_{cu} – specific heat of copper
c_{m} – specific heat of metal

c_{w}	– specific heat of tungsten
c^{p}	– material parameter in tensor function $\mathbf{M}(\mathbf{D})$
c^{r}	– material parameter in tensor function $\mathbf{L}(\mathbf{D})$
$\bar{c}$	– effective specific heat
C	– constant
$\widetilde{\mathbf{C}}$	– effective compliance tensor
$[\widetilde{\mathbf{C}}^{\mathrm{e}}]$	– effective tangent elastic compliance matrix
$[^{\mathrm{s}}\widetilde{\mathbf{C}}^{\mathrm{e}}]$	– effective secant elastic compliance matrix
$[\widetilde{\mathbf{C}}^{\mathrm{p}}]$	– effective tangent plastic compliance matrix
$[\widetilde{\mathbf{C}}^{\mathrm{epd}}]$	– effective tangent elasto-plastic-damage compliance matrix
$\widetilde{\mathbf{C}}_{\mathrm{ti}}$	– effective transversely isotropic compliance matrix
d	– material constant
d	– edge displacement
d_{inc}	– increment of edge displacement
$d_{<hkl>}$	– interplanar distance
d_0	– initial interplanar distance
$\mathcal{D}$	– bending stiffness
D_{eq}	– equivalent damage
D_{a}	– axial damage
D_{s}	– deviatoric damage variable
D_{t}	– transverse damage
D_{v}	– volumetric damage variable
$\mathbf{D}$	– damage tensor
$\mathbf{D}^{(e)}$	– element constitutive matrix
$\overline{\mathbf{D}}$	– nonlocally defined damage tensor
e	– Euler's number
E	– elastic modulus
E_0	– elastic modulus of homogeneous material
E_{a}	– axial elastic modulus
E_{c}	– elastic modulus of ceramic
E_{cu}	– elastic modulus of copper
E_{m}	– elastic modulus of metal
E_{t}	– transverse elastic modulus
E_{w}	– elastic modulus of tungsten
$\widetilde{E}^{\pm}$	– effective elastic modulus under tension and compression
$\mathcal{E}$	– energy
$\mathbf{E}$	– electric field vector
$\mathbf{E}$	– elastic constitutive (Hooke) tensor
$\widetilde{\mathbf{E}}^{\pm}$	– effective elastic tensor under tension and compression
f	– yield function
f	– volume fraction of reinforcement
$\overline{f}$	– nonlocal field function

$f_{1,2}$ – radius-dependent functions of axial displacement decomposition
F – dissipation potential
F – structure factor
F^{d} – damage part of dissipation potential
F^{p} – plastic part of dissipation potential
F_{D} – damage part of dissipation potential
g – auxiliary function of isotropic damage
G – shear modulus
G – fracture energy release rate
G – Green's function
$\overline{G}$ – effective shear modulus
G_{a} – axial shear modulus
G_{c} – critical fracture energy release rate
G_{t} – transverse shear modulus
h – thickness
h – Gauss distribution function
h – bell shape function
$h, \hbar$ – Planck's constants
h – crack closure parameter
h_{c} – critical value of crack closure parameter
$\mathbf{h}$ – crack closure tensor
H – Heaviside step function
$\mathbf{H}$ – damage force conjugate of $\boldsymbol{\chi}$
$[\mathbf{H}]$ – Hessian matrix
$[\overline{\mathbf{H}}]$ – nonlocal Hessian matrix
$\widetilde{\mathbf{H}}$ – effective Hill's plastic characteristic tensor
i – summation index
I – scattering intensity
I_0 – incident power per unit surface
I – index of eigenvalue
$\mathbf{I}$ – fourth rank unit tensor
$\mathbf{1}$ – second rank unit tensor
j – summation index
$\mathbf{J}$ – Jacobian matrix
$\widetilde{\mathbf{J}}$ – effective damage characteristic tensor

k – summation index
k – coefficient of modified elastic strain tensor
k – cuff-off factor
$k_{1,2}$ – constants of exponential type gradation
k – number of buckling mode
K – bulk modulus

$\overline{K}$ – effective bulk modulus

K_d – material constant

$\mathbf{k}^{(e)}$ – element stiffness matrix

$\mathbf{k}_0$ – wave vector

l – summation index

l – internal lenght

l_p – length of flexible cords

L – normalization constant

$\mathbf{L}$ – fourth-rank tensor describing influence of damage on damaged surface

m – mass

m – electron mass

$m_{1,2}$ – constants of power type gradation

m_N – neutron mass

M_r – radial bending moment

$\mathbf{m}$ – auxiliary second rank tensor for plastic strain evolution

$\mathbf{M(D)}$ – fourth rank damage effect tensor

N – number of cycles

$\mathbf{n}$ – unit vector

$N_{r,\varphi}$ – components of membrane stress resultants in polar coordinates

N_0 – shear force

p – pressure, loading

p – accumulated plastic strain

$\overline{p}$ – accumulated plastic strain at neck bottom

p_cr – critical value of load

p_D – damage threshold critical plastic strain

$\overline{P}$ – average property of two-phase material

P_cr – critical force

P_i – ith independent loading

P_i – critical value of ith independent loading

$P_{1,2}$ – properties of two-phase material constituents

$\mathbf{P}_\varepsilon^{+/-}$ – fourth rank projection operator of strain

$\mathbf{P}_\sigma^{+/-}$ – fourth rank projection operator of stress

q_v – inner heat source

q_z – axial component of body force

q_0 – exchanged heat

$\mathbf{q}$ – heat flux vector

Q – rate of internal heat source

$\overline{Q}_r$ – radial shear stress resultant

$\mathbf{Q}$ – scattering vector

r – radius

r – isotropic plastic hardening variable

r_0	– radius of electron
r_1	– radius of neck
r_1, r_2, r_3, r_4	– radii of brake disc
$\overline{r}$	– radius of specimen
$\widetilde{r}$	– effective isotropic plastic hardening variable
$\mathbf{r}$	– position vector
R	– interaction length
R	– thermodynamic force conjugate of r
$\widetilde{R}$	– effective thermodynamic force conjugate of r
R_0	– initial yield stress
$R_{0.2}$	– yield stress corresponding to plastic strain equal 0.2%
R_∞	– plastic material constant in Hayakawa–Murakami equation
R_g	– gyration radius of particle
s	– operator of neutron spin
s_{ij}	– deviatoric stress
$\overline{s}_{ij}$	– effective deviatoric stress
$\overline{s}_z$	– effective axial stress
S_s	– deviatoric damage strength material parameter
S_v	– volumetric damage strength material parameter
$\mathbf{S}$	– stiffness tensor
$\mathbf{S}$	– Poynting vector
$\widetilde{\mathbf{S}}$	– effective stiffness tensor
$[^\mathrm{s}\widetilde{\mathbf{S}}]$	– effective secant stiffness matrix
$^\mathrm{s}\overline{\widetilde{\mathbf{S}}}\left(\varepsilon^\mathrm{e}, \overline{\mathbf{D}}\right)$	– nonlocal secant stiffness
$\overline{\widetilde{\mathbf{S}}}\left(\varepsilon^\mathrm{e}, \overline{\mathbf{D}}\right)$	– nonlocal effective tangent stiffness tensor
t	– time
t_cr	– critical time
T	– temperature
T_f	– temperature of gas
T_ref	– reference temperature
T_∞	– temperature beyond boundary layer
u	– radial displacement
$\mathbf{u}$	– displacement vector
U	– potential energy
v	– hoop displacement
v_N	– neutron velocity
V	– volume
$V_\mathrm{A,B}$	– volume fractions of two-phase material constituents
V_N	– volume of brake fluid
w	– axial displacement
x	– coordinate

$\mathbf{X}$	– thermodynamic force conjugate of $\boldsymbol{\alpha}$
$\widetilde{\mathbf{X}}$	– thermodynamic effective force conjugate of $\boldsymbol{\alpha}$
y	– coordinate
Y_{s}	– deviatoric damage force conjugate
Y_{v}	– volumetric damage force conjugate
$\mathbf{Y}$	– thermodynamic force conjugate of $\mathbf{D}$
$\overline{\mathbf{Y}}$	– thermodynamic nonlocal force conjugate of $\overline{\mathbf{D}}$
z	– coordinate
Z	– atomic number
α	– angle
α	– coefficient of Hayhurst function
α	– thermal expansion coefficient
α_0	– thermal expansion coefficient of homogeneous material
$\alpha_1, \alpha_2, \alpha_3$	– direction cosines
α_{a}	– axial thermal expansion coefficient
α_{c}	– thermal expansion coefficient of ceramic
α_{cu}	– thermal expansion coefficient of copper
α_{m}	– thermal expansion coefficient of metal
α_{t}	– transverse thermal expansion coefficient
α_{w}	– thermal expansion coefficient of tungsten
$\widetilde{\alpha}$	– effective thermal expansion coefficient
$\boldsymbol{\alpha}$	– kinematic plastic hardening variable
$\widetilde{\boldsymbol{\alpha}}$	– effective kinematic plastic hardening variable
β	– material constant of exponential gradation
β	– coefficient of Newtonian convection
β	– isotropic damage hardening variable
$\overline{\beta}$	– nonlocal damage hardening variable
γ	– coefficient of Winkler foundation
γ	– material constant of exponential gradation
γ_{ij}	– shear strain
δ	– material constant of exponential gradation
δ_{ij}	– Kronecker's symbol
ε	– angular acceleration
ε	– uniaxial strain
$\overline{\varepsilon}$	– strain at bottom neck
$\varepsilon_{\langle hkl \rangle}$	– lattice strain
$\boldsymbol{\varepsilon}$	– strain tensor
$\boldsymbol{\varepsilon}^*$	– modified strain tensor for damage deactivation
$\boldsymbol{\varepsilon}^{\mathrm{e}}$	– elastic part of strain tensor
$\boldsymbol{\varepsilon}^{\mathrm{p}}$	– plastic strain tensor
$\boldsymbol{\varepsilon}^{*\mathrm{e}}$	– elastic part of modified strain tensor
ζ	– unilateral damage parameter

$\eta_1, \eta_2, \eta_3, \eta_4$ – material parameters of Murakami–Kamiya model
θ – difference between actual and reference temperature
θ – scattering angle
$\vartheta_1, \vartheta_2, \vartheta_3, \vartheta_4$ – material parameters of Hayakawa–Murakami model
$\imath$ – imaginary unit $= \sqrt{-1}$
κ – elastic force constant of electron
λ – thermal conductivity coefficient
λ – Lamé constant
λ – wave length
λ_c – thermal conductivity coefficient of ceramic
λ_{cu} – thermal conductivity coefficient of copper
λ_g – thermal conductivity of the gas
λ_m – thermal conductivity coefficient of metal
$\dot{\lambda}^d$ – damage multiplier
$\dot{\overline{\lambda}}^d$ – nonlocal damage multiplier
$\dot{\lambda}^p$ – plastic multiplier
λ_w – thermal conductivity coefficient of tungsten
λ_0 – thermal conductivity coefficient of homogeneous material
λ_{ij} – thermal conductivity matrix
$\widetilde{\lambda}$ – effective thermal conductivity coefficient
$\overline{\lambda}$ – plastic multiplier at bottom of neck
$\boldsymbol{\lambda}$ – thermal conductivity tensor
$\widetilde{\boldsymbol{\lambda}}$ – effective thermal conductivity tensor
μ – Lamé constant
ν – Poisson's ratio
ν – wave frequency
ν_a – axial Poisson's ratio
ν_t – transverse Poisson's ratio
ξ – dimensionless coordinate
$\boldsymbol{\xi}$ – position vector
ρ – scattering length density
ϱ – mass density
σ – uniaxial stress
σ – Thomson's scattering cross-section
σ_0 – material constant, reference stress
$\sigma_{r,\varphi,z}$ – components of stress tensor in cylindrical coordinate system
σ_{eq} – von Mises stress
σ_y – yield point stress
σ_a^u – axial tensile stress
σ_t^u – transverse tensile stress

$\boldsymbol{\sigma}$	– stress tensor
$\boldsymbol{\sigma}\prime$	– stress deviator
$\boldsymbol{\sigma}^*$	– modified stress tensor
$\boldsymbol{\sigma}_{\mathrm{macro}}$	– macrostress
$\boldsymbol{\sigma}_{\mathrm{tot}}^{\mathrm{i}}$	– total macrostress
$\boldsymbol{\sigma}_{\mathrm{mE}}^{\mathrm{i}}$	– elastic mismatch microstress
$\boldsymbol{\sigma}_{\mathrm{mT}}^{\mathrm{i}}$	– thermal mismatch microstress
$\boldsymbol{\sigma}_{\mathrm{tot}}^{\mathrm{reinf}}$	– macrostress of reinforcement
$\boldsymbol{\sigma}_{\mathrm{tot}}^{\mathrm{matrix}}$	– macrostress of matrix
$\overline{\sigma}_{r,t,z,\mathrm{i}}$	– stress components and stress intensity at bottom of neck
$\widetilde{\boldsymbol{\sigma}}$	– effective stress tensor
$\widetilde{\boldsymbol{\sigma}}^{\pm}$	– effective stress tensor referring to damage under tension and compression, respectively
τ	– time
τ_{rz}	– shear stress
ϕ	– time-wave function
φ	– angle
φ	– hoop coordinate
φ	– weight function
χ	– Hayhurst function
$\boldsymbol{\chi}$	– kinematic damage hardening variable
ψ	– angle
ψ	– space-wave function
ω	– angular velocity
Γ	– Gibb's thermodynamic potential
Γ^{d}	– damage part of Gibb's thermodynamic potential
Γ^{d}	– plastic part of Gibb's thermodynamic potential
Λ	– eigenvalue
$\boldsymbol{\Lambda}^{\mathrm{d}}$	– auxiliary second rank tensor for $\dot{\lambda}^{\mathrm{d}}$ representation
$\boldsymbol{\Lambda}^{\mathrm{P}}$	– auxiliary second rank tensor for $\dot{\lambda}^{\mathrm{P}}$ representation
Θ	– dilatancy
Σ	– macroscopic Thomson's scattering cross-section
$\Phi, \overline{\Phi}$	– non-negative dissipation function and functional associated to it
Φ_0	– photon flux
Ψ	– Helmholtz free energy
Ψ^{d}	– damage part of Helmholtz free energy
Ψ^{e}	– elastic part of Helmholtz free energy
Ψ^{P}	– plastic part of Helmholtz free energy
Ω_{d}	– domain of regularization
$\Omega^{(e)}$	– domain of finite element

1

Material Properties for High Temperature Applications

1.1 Technical Demands and Historical Outline

In many applications, especially in the space industry (e.g. jet engines and rockets) as well as electronic industry (e.g. superheat resistant materials for thermal protection, vehicle and personal body armour, electromagnetic sensors), structures or part of structures are exposed to high temperature, usually up to 2000K or even 3500K in some parts of rocket engines (Schulz *et al.* [202]), high temperature gradients, and/or cyclic temperature changes or impact loading. Conventional metallic materials, such as carbon steels or stainless steels: ASTM 321, ASTM 310, nickel- or aluminium-based alloys cannot resist such high temperatures (Odqvist [175]). The first method to improve the resistance of metallic structures against extreme temperature conditions consists in covering the structure (a substrate) with a ceramic layer (*Thermal Barrier Coating* – TBC), since ceramics are known for their high thermal resistance. For instance, in a metal–ceramic composite: Al–SiC the thermal conductivities ratio is approximately equal: $\lambda_m/\lambda_c = 3.6$, the thermal expansion coefficients ratio: $\alpha_m/\alpha_c = 5$, whereas the elastic moduli ratio: $E_m/E_c = 0.16$ (Poterasu *et al.* [181]). In the case of Ni–Al$_2$O$_3$ composite the corresponding ratios are: $\lambda_m/\lambda_c = 2.95$, $\alpha_m/\alpha_c = 1.51$ and $E_m/E_c = 0.50$ (Chen and Tong [44]). Hence, at the metal–ceramic interface, severe discontinuity of thermomechanical properties occurs, which results in high strain and stress mismatch at the interface; as a consequence, delamination or failure of the coating is rapidly observed. As a remedy to these disadvantages the concept of *Functionally Graded Materials* – FGM, was developed in Japan in the 1980s (Yamanouchi *et al.* [241]), giving structural components a spatial gradient in thermomechanical properties. The spatial gradient is achieved by use of two-component composites. The volume fraction of the composite constituents vary spatially such that the effective thermomechanical properties change smoothly from that of one material (e.g. ceramic) to the other (e.g. metal) (Fig. 1.1)

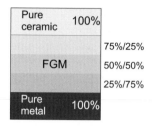

Fig. 1.1. Functionally graded ceramic–metal system (after Rangaraj *et al.* [189])

In this way, in the case of a TBC deposited on a metallic substrate, the heat-resistant ceramic layer and the solid metal are separated by functionally graded FG layers, the composition of which varies from pure ceramic to pure metal over a thickness of several 100 microns. The processing technologies for TBCs and FGMs lead to residual stresses, which are built-in during cool-down from the elevated fabrication temperature. These residual stresses may be significant relative to thermomechanical stresses applied subsequently. As regards FG layer processing, *Plasma Spray Thermal Barrier Coating* (PS–TBC) leads to lamellar microstructures, whereas columnar–lamellar microstructures are produced when using *Electron Beam Physical Vapour Deposition* (EBPVD) (Lee *et al.* [144], Schulz *et al.* [202]). Hence, the corresponding thermomechanical properties of FGMs are inhomogeneous and orthotropic rather than isotropic (Schulz *et al.* [202]).

A general review article on the application of the several ceramic materials to TBCs is given by Lee *et al.* [144]. Selected thermomechanical properties: elastic modulus E, thermal expansion coefficient α and thermal conductivity λ are summarized for frequently used ceramics for TBCs in Table 1.1.

Table 1.1. Selected thermomechanical properties of conventional TBC materials (after Lee *et al.* [144])

Coating Material	Elastic Modulus E[GPa]	Coeff. Th. Cond. λ[W/mK] at 1127°C	Th. Exp. Coeff. $\alpha[10^{-6}1/\text{K}]$
Partially Stabilized Zirconia $Y_2O_3+ZrO_2$	205	2.0	8.9–10.6
Al_2O_3	380–434	5.8	7.2–8.6
Mullite $3Al_2O_3 \cdot 2SiO_2$	145	3.3	5.7
TiO_2	283	3.3	9.4
$Ca_{0.5}Sr_{0.5}Zr_4P_6O_{24}$	70	1.0 (1000°C)	3.0

Zirconia ZrO_2 partially stabilized by Y_2O_3, deposited as a ceramic layer on a substrate material by various techniques, such as *Air Plasma Spray* (APS) or

EBPVD, is a commonly used ceramic coating because of its low thermal conductivity λ and a relatively high coefficient of thermal expansion α. However, its oxygen conductivity is rather high when compared to Al_2O_3-based alumina. Zirconia ZrO_2-based materials are therefore good materials for thermal insulation, but rather bad as an oxygen transport barrier. In contrast, alumina Al_2O_3-based ceramics are considered good materials for oxygen transport insulation, but their thermal insulation properties are not as good as zirconia. That is why more complex multifunctional coating systems are also in use, with two protecting ceramic layers against thermal and oxygen conductivity deposited on the substrate metallic material, and two additional FG interface layers introduced in order to mitigate the effects of sharp thermomechanical property mismatch at ceramic A–ceramic B–metal M interfaces (Fig. 1.2).

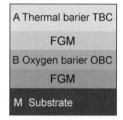

Fig. 1.2. Multilayer coating system for both thermal insulation and oxidation resistance (after Lee *at al.* [144])

1.2 Mechanical and Thermal Properties of Materials for High Temperature Service

Large scatter in the thermomechanical properties of materials used for thermal protection of metallic substrates is observed. It results from both the various chemical compositions and different manufacturing techniques used for production. Also the measurement techniques and the test conditions play en essential role. Among them, the temperature of the test strongly affects the material data.

A typical TBC system consists of a ceramic coating (e.g. zirconia or mullite) deposited over the FGM bond coat layer (e.g. intermetallic) on a substrate material (e.g. steel- or nickel-based superalloys). The most important mechanical and thermal properties of the constituent materials, such as: *elastic modulus, coefficient of thermal conductivity, thermal expansion coefficient* and *specific heat* are strongly temperature dependent constants: $E(T)$, $\lambda(T)$, $\alpha(T)$ and $c(T)$. Temperature dependence of other properties such as: *mass density* and *Poisson's ratio*, ρ and ν, is not important and can usually be ignored. The typical ceramic–intermetallic system is characterized in Table 1.2. (after:

Rangaraj and Kokini [187, 188, 189]). Data in brackets are taken from Kokini *et al.* [129].

Table 1.2. Temperature-dependent mechanical and thermal properties of zirconia-based and mullite-based TBCs (after: Rangaraj and Kokini [187, 188, 189])

Constituent Material	Temperature T[K]	Elastic Modulus E[GPa]	Thermal Conduct. λ[W/mK]	Thermal Expansion α[10^{-6}1/K]	Specific Heat c[J/molK]
Ceramic A–Zirconia	300	13.6 (36)	0.67 (0.5)	7.5 (8.0)	420
YSZ	1000	10.4	0.58	9.0	547
(Y_2O_3+ZrO_2)	1500	8.0	0.56	9.7	569
Ceramic B–Mullite	300	21 (30)	1.32 (1.3)	5.1 (4.5)	838
	1000	21	1.32	5.2	1202
($3Al_2O_3 \cdot 2SiC$)	1500	21	1.34	5.3	1219
Bond coat–BC	300	64	3.88	10.3	460
Intermetalic	1000	53.0	7.93	11.0	617
NiCoCrAlY	1500	43.0	9.86	11.4	620

Another set of data for constituent materials of TBC system: nickel-based superalloy substrate – NiCrAlY bond coat and the two-phase metal/ceramic (zirconia) FGM are reported in Yildrim and Erdogan [243]. Data in brackets refer to ZrO_2+Y_2O_3/NiCoCrAlY–TBC at room temperature according to Gan and Ng [78], Table 1.3.

Table 1.3. Temperature-dependent properties of zirconia–nickel-based TBC (after: Yildrim and Erdogan [243])

Constituent Material	Temperature T[K]	Elastic Modulus E[GPa]	Th. Expansion α[10^{-6}1/K]	Th. Conduct. λ[W/mK]
Zirconia–based	295	27.6 (53)	10.01 (7.2)	(1.5)
metal/ceramic	839	6.9	11.01	
FGM	1422	1.84	12.41	
Bond coat	295	137.9	15.16	
NiCrAlY	839	121.4	15.37	
	1422	93.8	17.48	
Substrate	295	175.8 (225)	13.91 (14)	(4.3)
Ni-based	839	150.4	15.36	
superalloy	1422	94.1	19.52	

Finally, for an alternative TBC system consisting of a zirconium oxide ZrO_2 ceramic layer and Ti–6Al–4V metal, the following data are detailed in

(A) Fujimoto and Noda [77], (B) Ootao et al. [173], (C) Noda et al. [165] Table 1.4.

Table 1.4. Properties of zirconia–titanium alloy at room temperature (after: Fujimoto and Noda [77], Ootao et al. [173], Noda et al. [165])

Constituent Material	Elastic Modulus E[GPa]	Th. Expansion $\alpha[10^{-6}1/\text{K}]$	Th. Conduct. λ[W/mK]	Specific Heat c[J/kgK]
Zirconia–based A,C	117.0	7.11	2.036	615.6
ceramic PSZ B	119.1	8.14	1.78	445
Titanium alloy A,C	66.2	10.3	18.1	808.3
Ti–6Al–4V B	107.3	8.91	5.74	526

Material properties of zirconia–titanium alloy TBC system change with temperature. The following empirical formulae can be quoted after Ootao et al. [173], for the zirconium oxide ceramic (ZrO_2) and titanium alloy (Ti–6Al–4V), applicable for the range of temperature $300 \text{ K} \leq T \leq 1300 \text{ } K$:
Ceramic

$$E_c = 132.2 - 50.3 \times 10^{-3}T - 8.1 \times 10^{-6}T^2[\text{GPa}]$$
$$\alpha_c = 13.3 \times 10^{-6} - 18.9 \times 10^{-9}T + 12.7 \times 10^{-12}T^2[1/\text{K}]$$
$$\lambda_c = 1.71 + 0.21 \times 10^{-3}T + 0.116 \times 10^{-6}T^2[\text{W/mK}] \qquad (1.1)$$
$$c_c = 2.74 \times 10^2 + 7.95 \times 10^{-1}T - 6.19 \times 10^{-4}T^2$$
$$+ 1.71 \times 10^{-7}T^3[\text{J/kgK}]$$

Metal

$$E_m = 122.7 - 0.0565T[\text{GPa}]$$
$$\alpha_m = \begin{cases} 7.43 \times 10^{-6} + 5.56 \times 10^{-9}T - 2.69 \times 10^{-12}T^2[1/\text{K}] \\ 10.291 \times 10^{-6}[1/\text{K}] \end{cases}$$
$$\lambda_m = 1.1 + 0.017T[\text{W/mK}] \qquad (1.2)$$
$$c_m = 3.5 \times 10^2 + 8.78 \times 10^{-1}T - 9.74 \times 10^{-4}T^2$$
$$+ 4.43 \times 10^{-7}T^3[\text{J/kgK}]$$

Alumina-based (Al_2O_3) ceramic materials exhibit lower oxygen diffusivity than zirconia-based (YSZ or PSZ) ceramics, but their thermal conductivity is much higher than that of zirconia. They are combined with nickel-based or aluminium-based alloys to produce the alternative Ni–Al_2O_3 or Al–Al_2O_3 TBCs. Comparison of mechanical and thermal properties for alumina Al_2O_3-based composites is presented in Table 1.5 (Chen and Tong [43], Cho and Shin [46] for Ni–Al_2O_3 and Wang et al. [232] for Al–Al_2O_3 composites).

Ni–Al_2O_3 composites or Ni–FGM–Al_2O_3 multi-layer composites are processed using a thermal spray method over the temperature range 20°C up to

Table 1.5. Comparison of properties of constituents of two alumina-based composites Ni–Al$_2$O$_3$ and Ti–Al$_2$O$_3$ (after: Chen and Tong [43],   Cho and Shin [46] for Ni–Al$_2$O$_3$ and Wang et al. [232] for Al–Al$_2$O$_3$ composites)

Composite Architecture	Elastic Modulus E[GPa]	Thermal Conductivity λ[W/mK]	Thermal Expansion Coefficient $\alpha[10^{-6}1/K]$	Specific Heat c[J/kgK]	Mass Density ρ[g/cm^3]
Ni	199.5	90.7	13.3	444	8.9
Al$_2$O$_3$	393.0	30.7	8.8	775	3.97
Al	73	154	23	963	2.8
Al$_2$O$_3$	380	46	8.5	765	3.96

800°C (Finot et al. [75]). The onset of cracking in alumina that deforms as a brittle solid at 1000°C, was analysed for cyclic thermal loading, concluding that at the Ni–FGM interface thermal cracking does not occur until after 750°C. That is why for higher temperature ranges, multilayer TBC systems are used, where the additional thermal resistant zirconia-based layer gives a better thermal insulation.

In general, following Herakovich and Aboudi [100], aluminium matrix composites have a usable temperature upward of 300°C, and titanium matrix composites can be used up to 800°C. For higher temperatures ceramic matrix materials have to be used (up to 2000°C). Properties of constituent materials of two titanium-based composites: titanium carbide–nickel (TiC–Ni) and titanium carbide–silicon carbide (TiC–SiC) are compared in Table 1.6. TiC–Ni composite is used to avoid the initiation of thermal fracture. The second, ceramic–ceramic composite TiC–SiC is a promising material because of good properties at high temperatures, especially corrosion and wear resistance. However, its brittleness and low failure resistance under thermal loads are basic disadvantages.

Table 1.6. Comparison of selected properties of two composites: ceramic–metal (TiC–Ni) and ceramic–ceramic (TiC–SiC) (after Wang et al. [233] for TiC–Ni and Jin and Paulino [119] for TiC–SiC)

Composite architecture	Elastic Modulus E[GPa]	Thermal Conductivity λ[W/mK]	Thermal Expansion Coefficient $\alpha[10^{-6}1/K]$	Specific Heat c[J/kgK]	Fracture Strength [MPa]
TiC	320	25.1	7.4	134	230
Ni	206	90.5	13.3	439.5	332
TiC	400	20	7.0	700	—
SiC	400	60	4.0	1000	—

In contrast to the previously discussed ceramic-based thermal barrier systems used for good thermal insulation, in some applications (e.g. thermonuclear reactor technology), actively cooled components that can withstand extremely high heat flux during transient and operating regimes, is the reason to use tungsten–copper (W–Cu) graded composite. The W–Cu system has excellent ability to reduce thermal stresses and very high thermal conductivity from the W surface to the Cu substrate (Ueda, [219, 220, 221, 222], Ueda and Gasik [223]) . The basic thermomechanical properties for W–Cu components measured at room temperature are listed in Table 1.7.

Table 1.7. Properties of W–Cu composite at room temperature (after: Ueda [219, 222])

Composite architecture	Elastic Modulus E[GPa]	Thermal Conductivity λ[W/mK]	Thermal Expansion Coefficient α[10^{-6}1/K]	Specific Heat c[J/kgK]	Mass Density ρ[g/cm^3]
W	414.4	182.9	4.6	128.2	1.93
Cu	128.0	384.3	17.0	378.5	8.93

Properties of tungsten and copper: E, λ and c are strongly temperature-dependent quantities, whereas α for tungsten essentially doesn't change with temperature. For temperature ranging from 293 K to 1300 K the following experimentally established formulae can be used (Ueda [219]):
Tungsten

$$E_w(T) = 423.4 - 0.044T\,[\text{GPa}]$$
$$\lambda_w(T) = 162.3 + 0.07T + 2.0 \times 10^{-5}T^2\,[\text{W/mK}] \quad (1.3)$$
$$\alpha_w(T) = 4.6 \times 10^{-6}\,[\text{1/K}]$$
$$c_w(T) = 10.011\sqrt{T} - 0.1362T\,[\text{J/kgK}]$$

Copper

$$E_{cu}(T) = 135.9 - 0.029T\,[\text{GPa}]$$
$$\lambda_{cu}(T) = 401.4 + 0.0625T\,[\text{W/mK}] \quad (1.4)$$
$$\alpha_{cu}(T) = 15.37 + 7.18 \times 10^{-3}T - 4.0 \times 10^{-6}T^2\,[10^{-6}\text{1/K}]$$
$$c_{cu}(T) = 36.675\sqrt{T} - 0.8333T\,[\text{J/kgK}]$$

In many cases composites have a directional structure, such that they exhibit different properties in different directions. In most engineering applications *unidirectional fibrous composites* are used with different axial properties (in the direction of fibres) and transverse properties (in the direction perpendicular to fibres). In the case of *orthotropic composites* the *effective composite stiffness matrix* referred to the principal material directions is given by

$$[\mathbf{S}] = \begin{bmatrix} S_{11} & S_{12} & S_{13} & 0 & 0 & 0 \\ S_{12} & S_{22} & S_{23} & 0 & 0 & 0 \\ S_{13} & S_{23} & S_{33} & 0 & 0 & 0 \\ 0 & 0 & 0 & S_{44} & 0 & 0 \\ 0 & 0 & 0 & 0 & S_{55} & 0 \\ 0 & 0 & 0 & 0 & 0 & S_{66} \end{bmatrix} \tag{1.5}$$

If plastic strains and damage are not considered, the linear Hooke's law holds

$$\{\boldsymbol{\sigma}\} = [\mathbf{S}]\{\boldsymbol{\varepsilon}\} \tag{1.6}$$

In a particular case of *transversely isotropic unidirectional material,* where direction (1) coincides with the fibres *axial directions* (a), whereas (2) and (3) are *transverse directions,* the number of independent mechanical moduli reduces to six: $E_a = E_{11}$, $E_t = E_{22} = E_{33}$, $G_a = G_{12} = G_{13}$, $G_t = G_{23}$, $\nu_a = \nu_{12} = \nu_{13}$ and $\nu_t = \nu_{23}$. Additionally, in case of thermal strains, two thermal expansion coefficients axial α_a and transverse α_t and two thermal conductivity coefficients in directions axial λ_a and transverse λ_t constitute thermal properties of the composite (Tamma and Avila [215])

$$\begin{Bmatrix} \varepsilon_{11} - \alpha_a \Delta T \\ \varepsilon_{22} - \alpha_t \Delta T \\ \varepsilon_{33} - \alpha_t \Delta T \\ \gamma_{23} \\ \gamma_{13} \\ \gamma_{12} \end{Bmatrix} = \begin{bmatrix} \dfrac{1}{E_a} & -\dfrac{\nu_a}{E_a} & -\dfrac{\nu_a}{E_a} & 0 & 0 & 0 \\ -\dfrac{\nu_a}{E_a} & \dfrac{1}{E_t} & -\dfrac{\nu_t}{E_t} & 0 & 0 & 0 \\ -\dfrac{\nu_a}{E_a} & -\dfrac{\nu_t}{E_t} & \dfrac{1}{E_t} & 0 & 0 & 0 \\ & & & \dfrac{1}{G_t} & 0 & 0 \\ & & & & \dfrac{1}{G_a} & 0 \\ & & & & & \dfrac{1}{G_a} \end{bmatrix} \begin{Bmatrix} \sigma_{11} \\ \sigma_{22} \\ \sigma_{33} \\ \sigma_{23} \\ \sigma_{13} \\ \sigma_{12} \end{Bmatrix} \tag{1.7}$$

In order to make a comparison between the properties of high-temperature composites and conventional engineering materials, let us examine the effect of temperature on the mechanical and thermal properties of a carbon steel, aluminium alloy and quartzitic concrete. Generally speaking, degradation of the mechanical properties elastic modulus E and yield stress σ_y is observed, accompanied by increasing values of the thermal properties thermal expansion coefficient α, thermal conductivity λ and specific heat c. Representative data are presented in Fig. 1.3a, b, c (mechanical properties) and Fig. 1.4a, b, c, d (thermal properties)

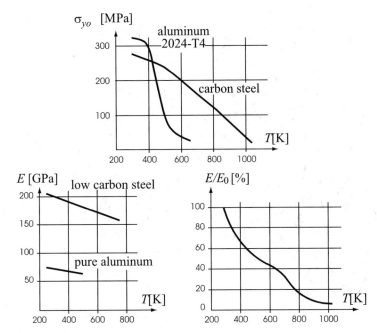

Fig. 1.3. Temperature dependence of mechanical properties of conventional materials (after Esbach and Sanders [67] (metals) and Schneider [198] (concrete))

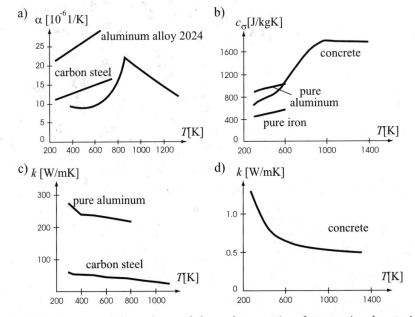

Fig. 1.4. Temperature dependence of thermal properties of conventional materials (after Esbach and Sanders [67] (metals) and Schneider [198] (concrete))

Table 1.8. Properties of selected unidirectional transversely isotropic composites (after: * Herakovich and Aboudi [100], ** Tamma and Avila, [215])

Composite material	T300/ 5280 *	Boron fibre/ Aluminium matrix *	Boron fibre/ Aluminium B/ Al6061-T0 **	Silicon carbide/ Titanium SCS-6/ Ti-15-3 *	Silicon carbide/ Titanium SCS-6/ Ti-15-3 **	Silicon carbide/ Aluminium SiC/ Al6061-T0 **
Density $\rho[\mathrm{g/cm^3}]$	1.54	2.65	$\nu_\mathrm{f}=47\%$	3.86	$\nu_\mathrm{f}=29\%$ $T=469°\mathrm{C}$	$\nu_\mathrm{f}=30\%$
Axial modulus $E_\mathrm{a}[\mathrm{GPa}]$	132	227	215.3– 217.1	221	171.0– 171.7	171.9– 172.4
Transverse modulus $E_\mathrm{t}[\mathrm{GPa}]$	10.8	139	131.9– 146.4	145	109.7– 118.1	104.0– 113.0
Poisson's ratio axial ν_a	0.24	0.24	0.172– 0.195	0.27	0.265– 0.278	0.251– 0.265
Poisson's ratio transverse ν_t	0.59	0.36	0.278– 0.317	0.40	0.338– 0.362	0.318– 0.341
Shear modulus axial $G_\mathrm{a}[\mathrm{GPa}]$	56.5	57.6	54.0– 75.6	53.2	42.5– 55.2	41.1– 54.9
Shear modulus transverse $G_\mathrm{t}[\mathrm{GPa}]$	33.8	49.1	50.0– 54.5	51.7	40.3– 42.6	38.8– 40.8
Axial tensile strength $\sigma_\mathrm{a}^\mathrm{u}[\mathrm{MPa}]$	1513	1290	—	1517	—	—
Transverse tensile strength $\sigma_\mathrm{t}^\mathrm{u}[\mathrm{MPa}]$	43.4	117	—	317	—	—
Thermal expansion coeff–axial $\alpha_\mathrm{a}[10^{-6}1/\mathrm{K}]$	-0.77	5.94	17.23– 17.81	6.15	10.71– 10.93	19.41– 19.84
Thermal expansion coefficient transverse $\alpha_\mathrm{t}[10^{-6}1/\mathrm{K}]$	25	16.6	17.81– 19.76	7.90	9.21– 9.99	18.08– 21.74

1.3 Macroscopic (Effective) Properties of Two-Phase Composites

Macroscopic (effective) mechanical and thermal properties of the two-phase composites depend on the volume fractions of both constituents and the physical properties of single materials at given temperature. The simplest *rule of mixture* is based on the linear approach, known as the *Voigt estimate* (Ueda [219, 222] for W–Cu composites, Shahsiah and Eslami [201] for steel–alumina composites, Elperin and Rudin [66] for W-C (tungsten carbide) ceramics and HS (steel) composite)

$$\bar{P}(T, \mathbf{x}) = P_1(T, \mathbf{x})V_1(\mathbf{x}) + P_2(T, \mathbf{x})V_2(\mathbf{x}) \tag{1.8}$$

An alternative rule, which determines the effective properties as a harmonic mean is called the *Reuss estimate*

$$\frac{1}{\bar{P}(T, \mathbf{x})} = \frac{V_1(\mathbf{x})}{P_1(T, \mathbf{x})} + \frac{V_2(\mathbf{x})}{P_2(T, \mathbf{x})} \tag{1.9}$$

However, the above simple equations have limited application since for volume fractions not close to 1 or 0 the Voigt estimate differs significantly from the Reuss estimate. To overcome this discrepancy a *modified rule of mixture* was proposed in Tomota *et al.* [218], where the empirical parameter called the *stress–strain transfer ratio q* is introduced to characterize uniaxial stress and strain states both in phase 1 (ceramic) and phase 2 (metal). Hence, for the *effective elasticity modulus $\bar{E}$* the following holds (Cho and Ha [45], Chen and Tong [43] for Ni–Al$_2$O$_3$)

$$\bar{E} = \frac{eV_\mathrm{m}E_\mathrm{m} + V_\mathrm{c}E_\mathrm{c}}{eV_\mathrm{m} + V_\mathrm{c}} \tag{1.10}$$

where

$$e = \frac{q + E_\mathrm{c}}{q + E_\mathrm{m}} \qquad q = \frac{\sigma_\mathrm{c} - \sigma_\mathrm{m}}{\varepsilon_\mathrm{m} - \varepsilon_\mathrm{c}} \qquad (0 < q < \infty) \tag{1.11}$$

The *effective thermal conductivity $\bar{\lambda}$* is given by a different formula (Jin and Paulino [119], Chen and Tong [43])

$$\bar{\lambda} = \lambda_\mathrm{m} + \frac{\lambda_\mathrm{m}V_\mathrm{c}(\lambda_\mathrm{c} - \lambda_\mathrm{m})}{\lambda_\mathrm{m} + (\lambda_\mathrm{m} - \lambda_\mathrm{c})V_\mathrm{m}/3} \tag{1.12}$$

whereas for the *effective coefficient of thermal expansion $\bar{\alpha}$* is

$$\bar{\alpha} = \alpha_\mathrm{m}\frac{(1/\overline{K} - 1/K_\mathrm{m})(\alpha_\mathrm{c} - \alpha_\mathrm{m})}{1/K_\mathrm{m} - 1/K_\mathrm{c}} \tag{1.13}$$

The *effective bulk modulus $\overline{K}$* and the *effective shear modulus $\overline{G}$* are:

$$\overline{K} = K_\mathrm{m} + \frac{aV_\mathrm{c}K_\mathrm{m}(K_\mathrm{c} - K_\mathrm{m})}{V_\mathrm{m}K_\mathrm{c} + aV_\mathrm{c}K_\mathrm{m}} \qquad \overline{G} = G_\mathrm{m} + \frac{bV_\mathrm{c}G_\mathrm{m}(G_\mathrm{c} - G_\mathrm{m})}{V_\mathrm{m}G_\mathrm{c} + bV_\mathrm{c}G_\mathrm{m}} \tag{1.14}$$

where

$$a = \frac{K_c(3K_m + 4G_m)}{K_m(3K_c + 4G_c)} \qquad b = \frac{(1+c)G_c}{G_m + cG_c} \qquad c = \frac{9K_m + 8G_m}{6K_m + 12G_m} \qquad (1.15)$$

For zirconia–titanium composite (ZrO_2/Ti–6Al–4V) Noda *et al.* [165] suggest the following rules of mixture:

$$\bar{\lambda} = \lambda_c \left[1 + \frac{3(\lambda_m - \lambda_c)V_m}{3\lambda_c + (\lambda_m - \lambda_c)V_c} \right]$$

$$\overline{E} = E_c \left[\frac{E_c + (E_m - E_c)V_m^{2/3}}{E_c + (E_m - E_c)(V_m^{2/3} - V_m)} \right]$$

$$\bar{\alpha} = \frac{\alpha_m V_m E_m/(1 - \nu_m) + \alpha_c V_c E_c/(1 - \nu_c)}{V_m E_m/(1 - \nu_m) + V_c E_c/(1 - \nu_c)}$$

$$\bar{c} = \frac{c_m \rho_m V_m + c_c \rho_c V_c}{\rho_m V_m + \rho_c V_c} \qquad (1.16)$$

Based on the micromechanics theory, more advanced formulae are derived for W–Cu FGM composite (Gasik and Lilius [86])

$$\overline{E} = E_{Cu} \left(1 + \frac{V_w}{\frac{E_w}{E_w - E_{Cu}} - \sqrt[3]{V_w}} \right)$$

$$\bar{\lambda} = \lambda_{Cu} \left(1 + \frac{V_w}{\frac{\lambda_w}{E_w - E_{Cu}} - \sqrt[3]{V_w}} \right) \qquad (1.17)$$

whereas for the effective value $\bar{\alpha}$ a more complicated rule is proposed.

Thermodynamics of Constitutive Modelling of Damaged Materials

2.1 Effect of Damage on the Mechanical and Thermal Properties of Materials

2.1.1 Heat Conduction in Damaged Materials

For nonhomogeneous materials the *heat balance equation* reads:

$$\text{div}\,[\mathbf{q}(\mathbf{x}, t)] + \frac{\partial}{\partial t}\,\{\rho(\mathbf{x}, t)c[\mathbf{x}, T(\mathbf{x}, t)]T(\mathbf{x}, t)\} - Q(\mathbf{x}, t) = 0 \qquad (2.1)$$

where $T(\mathbf{x}, t)$ is the *absolute temperature field*, $c[\mathbf{x}, T(\mathbf{x}, t)]$ is the *specific heat*, $\mathbf{q}(\mathbf{x}, t)$ is the *heat flux vector*, and $Q(\mathbf{x}, t)$ is the rate of *internal heat source*. The heat flux vector is given by the Fourier *heat conduction law*

$$\mathbf{q}(\mathbf{x}, t) = -\boldsymbol{\lambda}[\mathbf{x}, T(\mathbf{x}, t)] \cdot \mathbf{grad}\ T(\mathbf{x}, t) \qquad (2.2)$$

where $\boldsymbol{\lambda}[\mathbf{x}, T(\mathbf{x}, t)]$ is the second-rank temperature-dependent *tensor of thermal conductivity* for a nonhomogeneous material. If internal heat sources are neglected, the steady state heat flux equation for time-independent, *nonhomogeneous* but *isotropic material* holds (Tanigawa [216, 217]):

$$\frac{\partial}{\partial x}\left[\lambda(x, y, z)\frac{\partial T(x, y, z)}{\partial x}\right] + \frac{\partial}{\partial y}\left[\lambda(x, y, z)\frac{\partial T(x, y, z)}{\partial y}\right]$$
$$+ \frac{\partial}{\partial z}\left[\lambda(x, y, z)\frac{\partial T(x, y, z)}{\partial z}\right] = 0 \qquad (2.3)$$

In a more general case of *thermal anisotropy* described by the tensor of thermal conductivity λ_{ij}, the general heat flux equation for time-independent, nonhomogeneous, anisotropic material without heat sources is

$$\frac{\partial}{\partial x_i}\left[\lambda_{ij}(\mathbf{x})\frac{\partial T(\mathbf{x})}{\partial x_j}\right] = 0 \qquad (2.4)$$

where λ_{ij} is the symmetric second-rank *tensor of thermal conductivity*

$$\boldsymbol{\lambda} = \left\{ \begin{matrix} \lambda_{xx} & \lambda_{xy} & \lambda_{xz} \\ & \lambda_{yy} & \lambda_{yz} \\ & & \lambda_{zz} \end{matrix} \right\} \tag{2.5}$$

If the explicit representation is used, (2.4) reads

$$\frac{\partial}{\partial x}\left[\lambda_{xx}(x,y,z)\frac{\partial T(x,y,z)}{\partial x} + \lambda_{xy}(x,y,z)\frac{\partial T(x,y,z)}{\partial y}\right.$$

$$\left. + \lambda_{xz}(x,y,z)\frac{\partial T(x,y,z)}{\partial z}\right]$$

$$+\frac{\partial}{\partial y}\left[\lambda_{yx}(x,y,z)\frac{\partial T(x,y,z)}{\partial x} + \lambda_{yy}(x,y,z)\frac{\partial T(x,y,z)}{\partial y}\right.$$

$$\left. + \lambda_{yz}(x,y,z)\frac{\partial T(x,y,z)}{\partial z}\right]$$

$$+\frac{\partial}{\partial z}\left[\lambda_{zx}(x,y,z)\frac{\partial T(x,y,z)}{\partial x} + \lambda_{zy}(x,y,z)\frac{\partial T(x,y,z)}{\partial y}\right.$$

$$\left. + \lambda_{zz}(x,y,z)\frac{\partial T(x,y,z)}{\partial z}\right] = 0 \tag{2.6}$$

In the case of *damage-induced anisotropy* of thermal conductivity, the *effective conductivity tensor* coupled with damage is (Skrzypek and Ganczarski [206])

$$\tilde{\boldsymbol{\lambda}}(\mathbf{D}) = \lambda_0(\mathbf{1} - \mathbf{D}) \tag{2.7}$$

Hence, Eq. (2.4) takes the form

$$\frac{\partial}{\partial x_i}\left[\lambda_0(I_{ij} - D_{ij})\frac{\partial T}{\partial x_j}\right] = 0 \tag{2.8}$$

If the second-rank damage tensor $\mathbf{D}$ is expressed in principal directions that coincide with the principal directions of orthotropy (1,2,3)

$$\mathbf{D} = \left\{ \begin{matrix} D_{xx} & D_{xy} & D_{xz} \\ & D_{yy} & D_{yz} \\ & & D_{zz} \end{matrix} \right\} \rightarrow \left\{ \begin{matrix} D_{11} & 0 & 0 \\ & D_{22} & 0 \\ & & D_{33} \end{matrix} \right\} \tag{2.9}$$

tensor $\boldsymbol{\lambda}$ (2.5) becomes diagonal in form

$$\boldsymbol{\lambda} = \left\{ \begin{matrix} \lambda_0(1-D_{11}) & 0 & 0 \\ & \lambda_0(1-D_{22}) & 0 \\ & & \lambda_0(1-D_{33}) \end{matrix} \right\} \tag{2.10}$$

Hence, Eq. (2.8) reduces to the explicit form

$$\frac{\partial}{\partial x_1}\left[\lambda_0(\mathbf{x})(1-D_{11})\frac{\partial T}{\partial x_1}\right] + \frac{\partial}{\partial x_2}\left[\lambda_0(\mathbf{x})(1-D_{22})\frac{\partial T}{\partial x_2}\right]$$
$$+\frac{\partial}{\partial x_3}\left[\lambda_0(\mathbf{x})(1-D_{33})\frac{\partial T}{\partial x_3}\right] = 0 \tag{2.11}$$

Equation (2.11) holds for a steady-state temperature field in an *initially non-homogeneous*, isotropic material, in which damage-acquired thermal orthotropy occurs.

2.1.2 Effect of Damage on the Elastic Stiffness of Materials

The influence of damage on the deterioration of mechanical properties is governed by a fourth-rank tensor function $\mathbf{M}(\mathbf{D})$ with the second-rank damage tensor $\mathbf{D}$ as argument, also called the *damage effect tensor* (Chen and Chow [41], Skrzypek [204]).

When the damage effect tensor is used, the *effective* damage-influenced *stiffness* $\widetilde{\mathbf{S}}$ and *compliance* $\widetilde{\mathbf{C}}$ *tensors* are as follows:

$$\widetilde{\mathbf{S}} = \mathbf{M}^{-1}: \mathbf{S} : \mathbf{M}^{-\mathrm{T}} \text{ and } \widetilde{\mathbf{C}} = \mathbf{M}^{\mathrm{T}}: \mathbf{C} : \mathbf{M} \tag{2.12}$$

where $\mathbf{M}$, $\mathbf{M}^{-1}$ and $\mathbf{M}^{-\mathrm{T}}$ denote the damage effect tensor, its inverse and inverse transposed, respectively. Representation of the fourth-rank damage effect tensor $\mathbf{M}$ in terms of components of the second-rank damage tensor $\mathbf{D}$ is the key information necessary to describe *deterioration of the mechanical properties* due to damage. When the *matrix–vector notation* is used for the constitutive law of damaged material, the 6×6 matrix representation $[\mathbf{M}]_{ij}$ can be applied. If, additionally, damage tensor $\mathbf{D}$ is expressed in terms of its principal components, $D_{11} = D_1$, $D_{22} = D_2$, $D_{33} = D_3$, $D_{12} = D_{23} = D_{31} = 0$, the matrix $[\mathbf{M}]$ becomes diagonal in form:

$$[\mathbf{M}(\mathbf{D})] = \begin{bmatrix} M_{11} & 0 & 0 & 0 & 0 & 0 \\ & M_{22} & 0 & 0 & 0 & 0 \\ & & M_{33} & 0 & 0 & 0 \\ & & & M_{44} & 0 & 0 \\ & & & & M_{55} & 0 \\ & & & & & M_{66} \end{bmatrix} = [\mathbf{M}(\mathbf{D})]^{\mathrm{T}} \tag{2.13}$$

Hence, if *material orthotropy* holds (1.5), the *effective* (damage-influenced) *elastic stiffness matrix*, retains the characteristic symmetry property

$$\left[\widetilde{\mathbf{S}}(\mathbf{D})\right] = \left[\mathbf{M}^{-1}\right]\left[\mathbf{S}\right]\left[\mathbf{M}^{-\mathrm{T}}\right] = \begin{bmatrix} \widetilde{S}_{11} & \widetilde{S}_{12} & \widetilde{S}_{13} & 0 & 0 & 0 \\ & \widetilde{S}_{22} & \widetilde{S}_{23} & 0 & 0 & 0 \\ & & \widetilde{S}_{33} & 0 & 0 & 0 \\ & & & \widetilde{S}_{44} & 0 & 0 \\ & & & & \widetilde{S}_{55} & 0 \\ & & & & & \widetilde{S}_{66} \end{bmatrix} \tag{2.14}$$

In the particular case of *initial elastic isotropy*, three commonly used elastic-damage compliance matrices are written as (Chen and Chow [41], Skrzypek [204, 205]).

$$\left[\tilde{\mathbf{C}}_1\right] = [\mathbf{M}_1]\,[\mathbf{C}]\,[\mathbf{M}_1]$$

$$= \frac{1}{E}
\begin{bmatrix}
\dfrac{1}{(1-D_1)^2} & \dfrac{-\nu}{(1-D_1)(1-D_2)} & \dfrac{-\nu}{(1-D_1)(1-D_3)} & 0 & 0 & 0 \\[2mm]
0 & \dfrac{1}{(1-D_2)^2} & \dfrac{-\nu}{(1-D_2)(1-D_3)} & 0 & 0 & 0 \\[2mm]
0 & 0 & \dfrac{1}{(1-D_3)^2} & 0 & 0 & 0 \\[2mm]
0 & 0 & 0 & \dfrac{1+\nu}{(1-D_2)(1-D_3)} & 0 & 0 \\[2mm]
0 & 0 & 0 & 0 & \dfrac{1+\nu}{(1-D_1)(1-D_3)} & 0 \\[2mm]
0 & 0 & 0 & 0 & 0 & \dfrac{1+\nu}{(1-D_1)(1-D_2)}
\end{bmatrix}
\tag{2.15}$$

$$\left[\tilde{\mathbf{C}}_2\right] = [\mathbf{M}_2]\,[\mathbf{C}]\,[\mathbf{M}_2]$$

$$= \frac{1}{E}
\begin{bmatrix}
\dfrac{1}{(1-D_1)^2} & \dfrac{-\nu}{(1-D_1)(1-D_2)} & \dfrac{-\nu}{(1-D_1)(1-D_3)} & 0 & 0 & 0 \\[2mm]
0 & \dfrac{1}{(1-D_2)^2} & \dfrac{-\nu}{(1-D_2)(1-D_3)} & 0 & 0 & 0 \\[2mm]
0 & 0 & \dfrac{1}{(1-D_3)^2} & 0 & 0 & 0 \\[2mm]
0 & 0 & 0 & 0 & 0 & 0 \\[2mm]
0 & 0 & 0 & 0 & 0 & 0 \\[2mm]
0 & 0 & 0 & \dfrac{1+\nu}{\left(1-\frac{D_2+D_3}{2}\right)^2} & 0 & 0 \\[2mm]
 & & & 0 & \dfrac{1+\nu}{\left(1-\frac{D_1+D_3}{2}\right)^2} & 0 \\[2mm]
 & & & 0 & 0 & \dfrac{1+\nu}{\left(1-\frac{D_1+D_2}{2}\right)^2}
\end{bmatrix}
\tag{2.16}$$

$$\left[\widetilde{\mathbf{C}}_3\right] = [\mathbf{M}_3]\,[\mathbf{C}]\,[\mathbf{M}_3]$$

$$= \frac{1}{E}
\begin{bmatrix}
\frac{1}{(1-D_1)^2} & \frac{-\nu}{(1-D_1)(1-D_2)} & \frac{-\nu}{(1-D_1)(1-D_3)} & 0 & 0 & 0 \\[4pt]
0 & \frac{1}{(1-D_2)^2} & \frac{-\nu}{(1-D_2)(1-D_3)} & 0 & 0 & 0 \\[4pt]
0 & 0 & \frac{1}{(1-D_3)^2} & 0 & 0 & 0 \\[4pt]
0 & 0 & 0 & \frac{1+\nu}{4}\left(\frac{1}{1-D_2}+\frac{1}{1-D_3}\right)^2 & 0 & 0 \\[4pt]
0 & 0 & 0 & 0 & \frac{1+\nu}{4}\left(\frac{1}{1-D_1}+\frac{1}{1-D_3}\right)^2 & 0 \\[4pt]
0 & 0 & 0 & 0 & 0 & \frac{1+\nu}{4}\left(\frac{1}{1-D_1}+\frac{1}{1-D_2}\right)^2
\end{bmatrix}
\tag{2.17}$$

In a more general case of *transversely isotropic unidirectional composite material* (1.7), the effect of transversely isotropic damage $D_1 = D_\mathrm{a}$, $D_2 = D_3 = D_\mathrm{t}$ applied to (2.15) and (1.7) yields the following representation of the effective damage-influenced compliance matrix

$$[\widetilde{\mathbf{C}}_{\mathrm{ti}}] =
\begin{bmatrix}
\frac{1}{E_\mathrm{a}(1-D_\mathrm{a})^2} & \frac{-\nu_\mathrm{a}}{E_\mathrm{a}(1-D_\mathrm{a})(1-D_\mathrm{t})} & \frac{-\nu_\mathrm{a}}{E_\mathrm{a}(1-D_\mathrm{a})(1-D_\mathrm{t})} & 0 & 0 & 0 \\[4pt]
 & \frac{1}{E_\mathrm{t}(1-D_\mathrm{t})^2} & \frac{-\nu_\mathrm{t}}{E_\mathrm{t}(1-D_\mathrm{t})^2} & 0 & 0 & 0 \\[4pt]
 & & \frac{1}{E_\mathrm{t}(1-D_\mathrm{t})^2} & 0 & 0 & 0 \\[4pt]
 & & & \frac{1}{G_\mathrm{t}(1+D_\mathrm{t})^2} & 0 & 0 \\[4pt]
 & & & & \frac{1}{G_\mathrm{a}(1+D_\mathrm{a})(1+D_\mathrm{t})} & 0 \\[4pt]
 & & & & & \frac{1}{G_\mathrm{a}(1+D_\mathrm{a})(1+D_\mathrm{t})}
\end{bmatrix}
\tag{2.18}$$

2.2 Thermodynamic Framework for Constitutive Modelling of Elasto-Plastic-Damage Materials

Two general questions arise when a thermodynamically consistent framework is established for a description of elasto-(visco)-plasticity coupled with damage (Bielski *et al.* [22]):

(i) how to include damage to the thermodynamic potential (state coupling);
(ii) how to couple dissipation potential (dissipation coupling).

2.2.1 State Coupling

State coupling through the damage coupled thermodynamic potential (the Helmholtz or the Gibbs potential)

Case a: No state coupling

The state potential (the *Helmholtz free energy*) term Ψ^{p} associated with the plastic hardening variables α_j is not coupled with damage (Chaboche [39]):

$$\Psi = \Psi^{\mathrm{e}}(\varepsilon^{\mathrm{e}}, \mathbf{D}) + \Psi^{\mathrm{p}}(\alpha_j) \tag{2.19}$$

Case b: State coupling trough the new damage term Ψ^{d}

The new scalar state variable β is introduced as a measure of the cumulative damage in the damage state potential Ψ^{d} (Cordebois and Sidoroff [51], Zhu and Cescotto [245]):

$$\Psi = \Psi^{\mathrm{e}}(\varepsilon^{\mathrm{e}}, \mathbf{D}, T) + \Psi^{\mathrm{p}}(\alpha_j) + \Psi^{\mathrm{d}}(\beta) \tag{2.20}$$

When the *Gibbs thermodynamic potential* Γ is used instead of the Helmholtz free energy $\Gamma(\boldsymbol{\sigma}, r, \mathbf{D}, \beta) = \boldsymbol{\sigma} : \varepsilon^{\mathrm{e}} - \Psi$ (Hayakawa and Murakami [98]), the alternative formulation holds:

$$\Gamma = \Gamma^{\mathrm{e}}(\boldsymbol{\sigma}, \mathbf{D}) + \Gamma^{\mathrm{p}}(r) + \Gamma^{\mathrm{d}}(\beta) \tag{2.21}$$

In a more general case when both kinematic and isotropic hardening are admitted for both plasticity $(\boldsymbol{\alpha}, r)$ and damage $(\boldsymbol{\chi}, \beta)$ terms, the following holds (Voyiadjis and Deliktas [225], Abu Al–Rub and Voyiadjis [6]):

$$\Psi = \Psi^{\mathrm{e}}(\varepsilon^{\mathrm{e}}, \mathbf{D}) + \Psi^{\mathrm{p}}(\boldsymbol{\alpha}, r) + \Psi^{\mathrm{d}}(\boldsymbol{\chi}, \beta) \tag{2.22}$$

Direct hardening – damage state coupling

This approach consists in replacing classical variables $\varepsilon^{\mathrm{e}}, \boldsymbol{\alpha}$ and r in the Helmholtz potential functions by the effective (damage coupled) variables $\widetilde{\varepsilon}^{\mathrm{e}}(\mathbf{D}), \widetilde{\boldsymbol{\alpha}}(\mathbf{D})$ and $\widetilde{r}(\mathbf{D})$ to yield (Saanouni *et al.* [193])

$$\Psi = \Psi^{\mathrm{e}}(\widetilde{\varepsilon}^{\mathrm{e}}, T) + \Psi^{\mathrm{p}}(\widetilde{\boldsymbol{\alpha}}, \widetilde{r}) \tag{2.23}$$

2.2.2 Dissipation Coupling

Strong dissipation coupling

In this approach a *single plastic dissipation potential coupled with damage* is used, instead of two independent dissipation potential functions: plastic and damage in a more general case. *Strong dissipation coupling* is restricted only to a ductile-like damage in elastic-plastic materials, where both dissipation mechanisms: plasticity and damage, are governed by the single plastic multiplier λ^{P}. Again, three possibilities are met (Chaboche [37], Skrzypek, [205])

Case a: No plastic hardening–damage coupling: coupling only through the effective stress concept $\tilde{\sigma}$ (Benallal [20], Lemaitre [147])

Case b: Partly coupled plastic hardening–damage: coupling through the effective kinematic hardening $\tilde{\mathbf{X}}$ (Lemaitre and Chaboche [145])

Case c: Fully coupled plastic hardening–damage: coupling through two effective hardening variables $\tilde{\mathbf{X}}$ and $\tilde{R}$ (Chaboche [39], Saanouni [192])

For instance, when the J_2-type mixed plastic hardening model is used (Chaboche and Rousselier [38]):

$$F^{\mathrm{P}}(\boldsymbol{\sigma}, \mathbf{X}, R) = J_2(\boldsymbol{\sigma} - \mathbf{X}) - R - \sigma_{\mathrm{y}} = 0 \tag{2.24}$$

The above listed possibilities yield the equations

$$\begin{aligned}
F_{\mathrm{a}}^{\mathrm{P}}(\tilde{\boldsymbol{\sigma}}, \mathbf{X}, R) &= J_2(\tilde{\boldsymbol{\sigma}} - \mathbf{X}) - R - \sigma_{\mathrm{y}} = 0 \\
F_{\mathrm{b}}^{\mathrm{P}}(\tilde{\boldsymbol{\sigma}}, \tilde{\mathbf{X}}, R) &= J_2(\tilde{\boldsymbol{\sigma}} - \tilde{\mathbf{X}}) - R - \sigma_{\mathrm{y}} = 0 \\
F_{\mathrm{c}}^{\mathrm{P}}(\tilde{\boldsymbol{\sigma}}, \tilde{\mathbf{X}}, \tilde{R}) &= J_2(\tilde{\boldsymbol{\sigma}} - \tilde{\mathbf{X}}) - \tilde{R} - \sigma_{\mathrm{y}} = 0
\end{aligned} \tag{2.25}$$

where, in the case of a scalar damage parameter D, it holds that

$$J_2(\tilde{\boldsymbol{\sigma}} - \mathbf{X}) = \left[\frac{3}{2} \left(\frac{\sigma_{ij}}{1 - D} - X_{ij} \right) \left(\frac{\sigma_{ij}}{1 - D} - X_{ij} \right) \right]^{1/2} \tag{2.26}$$

Formulae (2.25) are used alternatively, however, only for case a, when approaching critical damage $D = 1$ the effective stress approaches zero $\tilde{\boldsymbol{\sigma}} = 0$. In case b, for critical damage only the isotropic hardening term tends to zero, whereas the kinematic one remains finite. In case c, both the isotropic and kinematic hardening terms remain finite at the critical damage.

Weak dissipation coupling

Strong dissipation coupling is frequently too restrictive in the sense that it does not allow for damage in the pure elastic range. On the other hand, in case of higher performance materials, especially composites, ceramics or functionally graded materials, the strong plasticity–damage coupling approaches are not applicable, and the more physically justified models based on the *weak plasticity–damage coupling* concept has to be used (Chow and Wang [47], Simo and Ju [203], Lemaitre and Chaboche [148], Chaboche [36], Hansen and Schreyer [94], Zhu and Cescotto [245], Hayakawa and Murakami [98, 99], Voyiadjis and Park [226, 227], Voyiadjis and Deliktas [225], Voyiadjis *et al.* [228]).

For instance, Chow and Wang [47] introduced in the stress space two independent dissipation potential functions, plastic F^{p} and damage F^{d}, as follows:

$$F^{\mathrm{p}} = \sigma^{\mathrm{p}}_{\mathrm{eq}} - [R_0 + R(r)] = 0 \qquad \sigma^{\mathrm{p}}_{\mathrm{eq}} = \sqrt{\tfrac{1}{2}\boldsymbol{\sigma}^T : \widetilde{\mathbf{H}} : \boldsymbol{\sigma}} \qquad (2.27)$$

$$F^{\mathrm{d}} = \sigma^{\mathrm{d}}_{\mathrm{eq}} - [B_0 + B(\beta)] = 0 \qquad \sigma^{\mathrm{d}}_{\mathrm{eq}} = \sqrt{\tfrac{1}{2}\boldsymbol{\sigma}^T : \widetilde{\mathbf{J}} : \boldsymbol{\sigma}}$$

Symbols $\widetilde{\mathbf{H}}$ and $\widetilde{\mathbf{J}}$ denote the *effective Hill's plastic characteristic* and the *effective damage characteristic* fourth-rank *tensors*, respectively, whereas $R(r)$ and $B(\beta)$ are respective *force conjugates* of plasticity and damage scalar hardening variables r and β.

A similar concept was used by Hayakawa and Murakami [98], who assumed the *dissipation potential* to be composed of two terms, *plastic* and *damage*, in the form:

$$F^{\mathrm{p}}(\boldsymbol{\sigma},R,\mathbf{D}) = \sigma_{\mathrm{eq}} - [R_0 + R(r)] = 0 \qquad (2.28)$$

$$\sigma_{\mathrm{eq}} = \sqrt{\tfrac{3}{2}\boldsymbol{\sigma}' : \mathbf{M}(\mathbf{D}) : \boldsymbol{\sigma}'}$$

$$F^{\mathrm{d}}(\mathbf{Y},B;\mathbf{D},r,\beta) = Y_{\mathrm{eq}} + c^r r \mathrm{Tr}\mathbf{D}\mathrm{Tr}\mathbf{Y} - [B_0 + B(\beta)] = 0$$

$$Y_{\mathrm{eq}} = \sqrt{\tfrac{1}{2}\mathbf{Y} : \mathbf{L}(\mathbf{D}) : \mathbf{Y}}$$

Symbols $\mathbf{M}(\mathbf{D})$ and $\mathbf{L}(\mathbf{D})$ denote fourth-rank tensor functions linear in $\mathbf{D}$, $\mathbf{Y}$ denotes force conjugate of $\mathbf{D}$, $R(r)$ and $B(\beta)$ are respective force conjugate of isotropic hardening variables for plasticity and damage r and β. The above equations describe the plastic dissipation surface and the damage dissipation surface (2.28), both defined in the stress space $\mathbf{Y}(\boldsymbol{\sigma})$ when the Gibbs potential is used as the state potential (Fig. 2.1). Note that both dissipation functions F^{p} and F^{d} are coupled via the damage variable $\mathbf{D}$, this being the common argument of both tensor functions $\mathbf{M}(\mathbf{D})$ and $\mathbf{L}(\mathbf{D})$. Additionally, it is assumed that the damage function F^{d} is affected by the plastic hardening variable r. On the other hand, the representation of the dissipation functions F^{p} and F^{d} (2.28) is restricted to the case of isotropic plasticity and damage hardening, r and β, exclusively.

The tensor functions $\mathbf{M}(\mathbf{D})$ and $\mathbf{L}(\mathbf{D})$ are specified in such a way, that in no-damage states both the yield and damage functions are isotropic

$$[\mathbf{M}(\mathbf{D})]_{ijkl} = \frac{1}{2}\left(\delta_{ik}\delta_{jl} + \delta_{il}\delta_{jk}\right)$$

$$+\frac{1}{2}c^{\mathrm{p}}\left(\delta_{ik}D_{jl} + D_{ik}\delta_{jl} + \delta_{il}D_{jk} + D_{il}\delta_{jk}\right)$$

$$[\mathbf{L}(\mathbf{D})]_{ijkl} = \frac{1}{2}\left(\delta_{ik}\delta_{jl} + \delta_{il}\delta_{jk}\right)$$ (2.29)

$$+\frac{1}{2}c^{\mathrm{r}}\left(\delta_{ik}D_{jl} + D_{ik}\delta_{jl} + \delta_{il}D_{jk} + D_{il}\delta_{jk}\right)$$

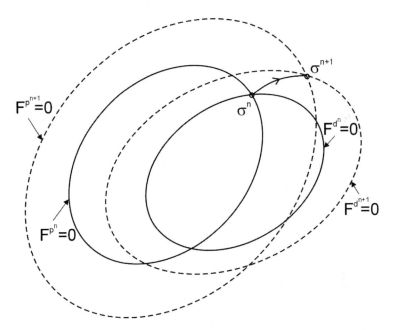

Fig. 2.1. Schematic illustration of a double-active dissipation process, plastic and damage

In a more general case when mixed (isotropic and kinematic) hardening is used for plasticity and damage, see (2.27) (Abu Al-Rub and Voyiadjis [6]), and the fully-coupled plastic hardening-damage through two effective plastic hardening variables $\widetilde{\mathbf{X}}$ and $\widetilde{R}$ is adopted (2.25), the following formulae are used (Abu Al-Rub and Voyiadjis [6]):

$$F^{\mathrm{P}}\left(\boldsymbol{\sigma}, \mathbf{X}, R; \mathbf{D}\right) = \sigma_{\mathrm{eq}}\left(\widetilde{\boldsymbol{\sigma}} - \widetilde{\mathbf{X}}\right) - \left[R_0 + \widetilde{R}(r)\right] = 0$$ (2.30)

$$\sigma_{\mathrm{eq}}\left(\widetilde{\boldsymbol{\sigma}} - \widetilde{\mathbf{X}}\right) = \sqrt{\frac{3}{2}\left(\widetilde{\boldsymbol{\sigma}} - \widetilde{\mathbf{X}}\right):\left(\widetilde{\boldsymbol{\sigma}} - \widetilde{\mathbf{X}}\right)}$$

$$F^{\mathrm{d}}\left(\mathbf{Y}, \mathbf{H}, B\right) = Y_{\mathrm{eq}}\left(\mathbf{Y} - \mathbf{H}\right) + c\mathbf{H} : \mathbf{H} - \left[B_0 + B(\beta)\right] = 0$$
$$Y_{\mathrm{eq}}\left(\mathbf{Y} - \mathbf{H}\right) = \sqrt{\left(\mathbf{Y} - \mathbf{H}\right) : \left(\mathbf{Y} - \mathbf{H}\right)}$$

where the effective variables $\widetilde{\boldsymbol{\sigma}}$ and $\widetilde{\mathbf{X}}$ are defined as follows

$$\widetilde{\boldsymbol{\sigma}} = \mathbf{M}(\mathbf{D}) : \boldsymbol{\sigma}, \qquad \widetilde{\mathbf{X}} = \mathbf{M}(\mathbf{D}) : \mathbf{X}, \qquad \widetilde{R} = \frac{R}{1 - D_{\mathrm{eq}}} \qquad (2.31)$$

and D_{eq} denotes the equivalent damage (Voyiadjis and Park [226])

$$D_{\mathrm{eq}} = \sqrt{\mathbf{D} : \mathbf{D}} \qquad (2.32)$$

2.2.3 State and Dissipation Equations for Elasto-Plasticity Coupled with Damage

State and dissipation potentials play a crucial role for the thermodynamically consistent formulation of the state and evolution equations. It is shown schematically in Fig. 2.2.

Applying the *state potential*, the *Helmholtz* $\Psi(\varepsilon^{\mathrm{e}}, r, \mathbf{D}, \beta)$ or the *Gibbs* $\Gamma(\boldsymbol{\sigma}^{\mathrm{e}}, r, \mathbf{D}, \beta)$, and following the scheme for irreversible thermodynamics, the elastic strain–stress equation of damaged materials are obtained (*state equations*) as

$$\boldsymbol{\sigma} = \frac{\partial \Psi^{\mathrm{e}}}{\partial \varepsilon^{\mathrm{e}}} = {}^{\mathrm{s}}\widetilde{\mathbf{S}}(\mathbf{D}) : \varepsilon^{\mathrm{e}} \quad \text{or} \quad \{\boldsymbol{\sigma}\} = \left[{}^{\mathrm{s}}\widetilde{\mathbf{S}}(\mathbf{D})\right]\{\varepsilon^{\mathrm{e}}\} \qquad (2.33)$$

or

$$\varepsilon^{\mathrm{e}} = \frac{\partial \Gamma^{\mathrm{e}}}{\partial \boldsymbol{\sigma}} = {}^{\mathrm{s}}\widetilde{\mathbf{C}}(\mathbf{D}) : \boldsymbol{\sigma} \quad \text{or} \quad \{\varepsilon^{\mathrm{e}}\} = \left[{}^{\mathrm{s}}\widetilde{\mathbf{C}}(\mathbf{D})\right]\{\boldsymbol{\sigma}\} \qquad (2.34)$$

where ${}^{\mathrm{s}}\widetilde{\mathbf{S}}(\mathbf{D})$ or ${}^{\mathrm{s}}\widetilde{\mathbf{C}}(\mathbf{D})$ denote the *effective elastic secant stiffnesses* or *elastic secant compliance* tensors, whereas $[{}^{\mathrm{s}}\widetilde{\mathbf{S}}(\mathbf{D})]$ or $[{}^{\mathrm{s}}\widetilde{\mathbf{C}}(\mathbf{D})]$ stand for the corresponding *secant stiffness* and *compliance matrices*, if the matrix–vector notation is used for damaged material (2.12). The thermodynamic force-conjugates of state variables $r, \mathbf{D}$ and β are obtained in a similar fashion.

In a simple case of *isotropic plasticity* and *damage hardening* we arrive at (Murakami and Kamiya [162])

$$R = \rho \frac{\partial \Psi^{\mathrm{p}}}{\partial r}, \qquad \mathbf{Y} = -\rho \frac{\partial \Psi^{\mathrm{e}}}{\partial \mathbf{D}}, \qquad B = \rho \frac{\partial \Psi^{\mathrm{d}}}{\partial \beta} \qquad (2.35)$$

or when the Gibbs formulation is used (Hayakawa and Murakami, 1997)

$$R = \rho \frac{\partial \Gamma^{\mathrm{p}}}{\partial r}, \qquad \mathbf{Y} = \rho \frac{\partial \Gamma^{\mathrm{e}}}{\partial \mathbf{D}}, \qquad B = \rho \frac{\partial \Gamma^{\mathrm{d}}}{\partial \beta} \qquad (2.36)$$

In a more general case, when both *isotropic* and *kinematic hardening* are applied for *plasticity* and *damage* (2.22), five thermodynamic force-conjugates

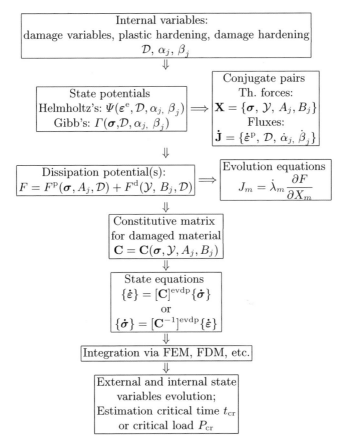

Fig. 2.2. Continuum damage mechanics CDM formulation based on the irreversible thermodynamics with internal variables (after Skrzypek [205])

of the state variables $\boldsymbol{\alpha}, r, \boldsymbol{\chi}, \beta$ and $\mathbf{D}$ are defined (Abu Al-Rub and Voyiadjis [6])

$$\mathbf{X} = \rho\frac{\partial \Psi^{\mathrm{p}}}{\partial \boldsymbol{\alpha}}, \quad R = \rho\frac{\partial \Psi^{\mathrm{p}}}{\partial r}, \quad \mathbf{H} = \rho\frac{\partial \Psi^{\mathrm{d}}}{\partial \boldsymbol{\chi}}, \quad B = \rho\frac{\partial \Psi^{\mathrm{d}}}{\partial \beta}, \quad \mathbf{Y} = -\rho\frac{\partial \Psi^{\mathrm{e}}}{\partial \mathbf{D}} \quad (2.37)$$

When, for instance, definitions of conjugate variables (2.35) are used, the *Clausius–Duhem inequality* for the non-negative dissipation due to plasticity and damage holds

$$\Phi = \boldsymbol{\sigma} : \dot{\boldsymbol{\varepsilon}}^{\mathrm{p}} - R\dot{r} - \mathbf{Y} : \dot{\mathbf{D}} - B\dot{\beta} \geq 0 \quad (2.38)$$

with two constraints $F^{\mathrm{p}} = 0$ and $F^{\mathrm{d}} = 0$ (2.28), if the concept of Hayakawa and Murakami [98] or Zhu and Cescotto [245] is used.

Hence, when two Lagrange multipliers $\dot{\lambda}^{\mathrm{p}}$ and $\dot{\lambda}^{\mathrm{d}}$ are introduced, the functional $\bar{\Phi}$ is defined as

$$\bar{\varPhi} = \varPhi - \dot{\lambda}^{\mathrm{P}} F^{\mathrm{p}} - \dot{\lambda}^{\mathrm{d}} F^{\mathrm{d}} = \boldsymbol{\sigma} : \dot{\boldsymbol{\varepsilon}}^{\mathrm{P}} - R\dot{r} - \mathbf{Y} : \dot{\mathbf{D}} - B\dot{\beta} - \dot{\lambda}^{\mathrm{P}} F^{\mathrm{p}} - \dot{\lambda}^{\mathrm{d}} F^{\mathrm{d}} \quad (2.39)$$

Maximizing next $\bar{\varPhi}$ we arrive at the plasticity and damage constitutive equations (*evolution equations* or *generalized normality rules*)

$$
\begin{aligned}
\frac{\partial \bar{\varPhi}}{\partial \boldsymbol{\sigma}} &= 0 & \rightarrow \dot{\boldsymbol{\varepsilon}}^{\mathrm{P}} &= \dot{\lambda}^{\mathrm{P}} \frac{\partial F^{\mathrm{p}}}{\partial \boldsymbol{\sigma}} \\
\frac{\partial \bar{\varPhi}}{\partial (-R)} &= 0 \rightarrow \dot{r} &= -\dot{\lambda}^{\mathrm{P}} \frac{\partial F^{\mathrm{p}}}{\partial R} \\
\frac{\partial \bar{\varPhi}}{\partial \mathbf{Y}} &= 0 & \rightarrow \dot{\mathbf{D}} &= -\dot{\lambda}^{\mathrm{d}} \frac{\partial F^{\mathrm{d}}}{\partial \mathbf{Y}} \\
\frac{\partial \bar{\varPhi}}{\partial (-B)} &= 0 \rightarrow \dot{\beta} &= -\dot{\lambda}^{\mathrm{d}} \frac{\partial F^{\mathrm{d}}}{\partial B}
\end{aligned}
\quad (2.40)
$$

If, on the other hand, a more general formulation of *isotropic* and *kinematic* *plasticity* and *damage hardening* (2.37) is used, we arrive at

$$
\begin{aligned}
\dot{\boldsymbol{\varepsilon}}^{\mathrm{P}} &= \dot{\lambda}^{\mathrm{P}} \frac{\partial F^{\mathrm{p}}}{\partial \boldsymbol{\sigma}} & \dot{\boldsymbol{\alpha}} &= -\dot{\lambda}^{\mathrm{P}} \frac{\partial F^{\mathrm{p}}}{\partial \mathbf{X}} & \dot{r} &= -\dot{\lambda}^{\mathrm{P}} \frac{\partial F^{\mathrm{p}}}{\partial R} \\
\dot{\mathbf{D}} &= -\dot{\lambda}^{\mathrm{d}} \frac{\partial F^{\mathrm{d}}}{\partial \mathbf{Y}} & \dot{\boldsymbol{\chi}} &= -\dot{\lambda}^{\mathrm{d}} \frac{\partial F^{\mathrm{d}}}{\partial \mathbf{H}} & \dot{\beta} &= -\dot{\lambda}^{\mathrm{d}} \frac{\partial F^{\mathrm{d}}}{\partial B}
\end{aligned}
\quad (2.41)
$$

If the above equations hold the *maximum dissipation principle* is satisfied, which states: "Actual state of the thermodynamic forces is that which maximizes the dissipation function over all other possible admissible states" (Hansen and Schreyer [94]). The *Khun–Tucker loading/unloading conditions* for both independent dissipation mechanisms, plasticity and damage, hold

$$\dot{\lambda}^{\mathrm{P}} \geq 0, \quad F^{\mathrm{p}} \leq 0, \quad \dot{\lambda}^{\mathrm{P}} F^{\mathrm{p}} = 0 \quad \text{and} \quad \dot{\lambda}^{\mathrm{d}} \geq 0, \quad F^{\mathrm{d}} \leq 0, \quad \dot{\lambda}^{\mathrm{d}} F^{\mathrm{d}} = 0 \quad (2.42)$$

Note that the plasticity–damage evolution processes (2.40) or (2.42) are history-dependent, hence the integration of the constitutive equations at each incremental step, must be performed from the last equilibrium step instead of simple cumulation of increments of stresses. In other words, at each integration point it is assumed that the state variables $(\{\boldsymbol{\varepsilon}\}, \{\boldsymbol{\alpha}\}, r, \{\mathbf{D}\}, \{\boldsymbol{\chi}\}, \beta)$ and their conjugate forces $(\{\boldsymbol{\sigma}\}, \{\mathbf{X}\}, R, \{\mathbf{Y}\}, \{\mathbf{H}\}, B)$ are known at the beginning of a time increment. In the frame of the *small strain theory* the strain increment is the sum of the elastic and plastic part

$$\{\dot{\boldsymbol{\varepsilon}}\} = \{\dot{\boldsymbol{\varepsilon}}^{\mathrm{e}}\} + \{\dot{\boldsymbol{\varepsilon}}^{\mathrm{P}}\} \quad (2.43)$$

whereas the stress increment $\{\dot{\boldsymbol{\sigma}}\}$ is sought along with conjugate pairs' increase of $\{\dot{\mathbf{X}}\}$ and $\{\dot{\boldsymbol{\alpha}}\}$, $\dot{R}$ and $\dot{r}$, $\{\dot{\mathbf{Y}}\}$ and $\{\dot{\mathbf{D}}\}$, $\{\dot{\mathbf{H}}\}$ and $\{\dot{\boldsymbol{\chi}}\}$, B and $\dot{\beta}$.

Hence, we need the incremental form of the elasto-damage constitutive equations (2.25)

$$\{\dot{\boldsymbol{\varepsilon}}^{\mathrm{e}}\} = \left[\widetilde{\mathbf{C}}^{\mathrm{e}}(\boldsymbol{\sigma}, \mathbf{D}) \right] \{\dot{\boldsymbol{\sigma}}\} \quad (2.44)$$

where $\widetilde{\mathbf{C}}^{\mathrm{e}}$ denotes the local *tangent compliance matrix* that has to be determined by the use of corresponding *secant compliance matrix* $^{\mathrm{s}}\widetilde{\mathbf{C}}^{\mathrm{e}}$ as follows (Bielski *et al.* [22]):

$$\widetilde{\mathbf{C}}^{\mathrm{e}}(\boldsymbol{\sigma},\mathbf{D}) = {}^{\mathrm{s}}\widetilde{\mathbf{C}}^{\mathrm{e}}(\mathbf{D}) + \frac{\partial\, {}^{\mathrm{s}}\widetilde{\mathbf{C}}^{\mathrm{e}}(\mathbf{D})}{\partial\mathbf{D}} : \frac{\partial\mathbf{D}}{\partial\boldsymbol{\sigma}} : \boldsymbol{\sigma} \qquad (2.45)$$

Finally, combining (2.43) and (2.44) we arrive at the coupled *incremental elasto-plastic-damage* constitutive equations

$$\{\dot{\boldsymbol{\varepsilon}}\} = \left[\widetilde{\mathbf{C}}^{\mathrm{e}}\right]\{\dot{\boldsymbol{\sigma}}\} + \left[\widetilde{\mathbf{C}}^{\mathrm{p}}\right]\{\dot{\boldsymbol{\sigma}}\} = \left[\widetilde{\mathbf{C}}^{\mathrm{epd}}(\boldsymbol{\sigma},\mathbf{D},\mathbf{Y},r)\right]\{\dot{\boldsymbol{\sigma}}\} \qquad (2.46)$$

The plastic compliance matrix is obtained from the plasticity–damage evolution equations (2.40) or (2.41) with plasticity and damage multipliers $\dot{\lambda}^{\mathrm{p}}$ and $\dot{\lambda}^{\mathrm{d}}$ determined from the consistency conditions for the yield and damage surface $\dot{F}^{\mathrm{p}} = 0$ and $\dot{F}^{\mathrm{d}} = 0$. Note that elements of the local tangent compliance or stiffness matrices are not only functions of damage variable $\mathbf{D}$ (2.15–2.18), but also depend on the conjugate forces $\boldsymbol{\sigma}$ and $\mathbf{Y}$. The initial compliance matrix $^{\mathrm{s}}\mathbf{C}$, the secant compliance matrix of damaged material $^{\mathrm{s}}\widetilde{\mathbf{C}}$, and the tangent compliance matrix $\widetilde{\mathbf{C}}$ of elasto-plastic-damage material are sketched in Fig. 2.3

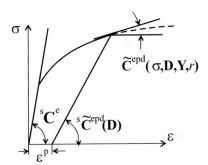

Fig. 2.3. Comparison of the initial effective-secant and effective-tangent compliances of elastic-plastic-damage material (Skrzypek [205])

In the particular case if *material orthotropy* holds, the direction of which coincides with damage principal directions, the incremental elasto-plastic-damage constitutive equations (2.46) retain the symmetry property:

$$\begin{Bmatrix} \dot{\varepsilon}_{11} - \tilde{\alpha}_{11}\dot{T} \\ \dot{\varepsilon}_{22} - \tilde{\alpha}_{22}\dot{T} \\ \dot{\varepsilon}_{33} - \tilde{\alpha}_{33}\dot{T} \\ \dot{\gamma}_{23} \\ \dot{\gamma}_{13} \\ \dot{\gamma}_{12} \end{Bmatrix} = \begin{bmatrix} \tilde{C}_{11} & \tilde{C}_{12} & \tilde{C}_{13} & 0 & 0 & 0 \\ & \tilde{C}_{22} & \tilde{C}_{23} & 0 & 0 & 0 \\ & & \tilde{C}_{33} & 0 & 0 & 0 \\ & & & \tilde{C}_{44} & 0 & 0 \\ & & & & \tilde{C}_{55} & 0 \\ & & & & & \tilde{C}_{66} \end{bmatrix} \begin{Bmatrix} \dot{\sigma}_{11} \\ \dot{\sigma}_{22} \\ \dot{\sigma}_{33} \\ \dot{\sigma}_{23} \\ \dot{\sigma}_{13} \\ \dot{\sigma}_{12} \end{Bmatrix} \qquad (2.47)$$

Note, that symbols $\tilde{C}_{ij}$ denote elements of the *effective tangent elasto-plastic-damage compliance matrix* and should read here as

$$\tilde{C}_{ij} \equiv \left[\tilde{\mathbf{C}}^{\rm epd}(\boldsymbol{\sigma}, \mathbf{D}, \mathbf{Y}, r, T) \right]_{ij} \tag{2.48}$$

whereas $\tilde{\alpha}_{ii}$ denote temperature-dependent principal components of the *effective thermal expansion coefficients tensor* $\tilde{\boldsymbol{\alpha}}(\mathbf{D}; T)$

The particular form of the matrix $[\tilde{\mathbf{C}}^{\rm epd}]$ depends not only on the state variables and the conjugate forces but, also, on a possible process characterization. Namely, it depends on whether the process on the incremental loading step is 'active' or 'passive' in a generalized sense for both plasticity and damage.

Table 2.1. Limit surfaces evolution $F^{\rm p}$ and $F^{\rm d}$ on the incremental loading step

Incremental loading		Dissipation surfaces		Effective compliance
plastic flow	damage	plastic flow	damage	matrix
no	no	$F^{\rm p} < 0$ or $F^{\rm p} = 0$ and $\dot{\lambda}^{\rm P} = 0$	$F^{\rm d} < 0$ or $F^{\rm d} = 0$ and $\dot{\lambda}^{\rm d} = 0$	$\mathbf{C}^{\rm epd} = \mathbf{C}^{\rm ed}$
no	yes	$F^{\rm p} < 0$ or $F^{\rm p} = 0$ and $\dot{\lambda}^{\rm P} = 0$	$F^{\rm d} > 0$ $(\dot{\lambda}^{\rm d} > 0)$	
yes	no	$F^{\rm p} > 0$ $(\dot{\lambda}^{\rm P} > 0)$	$F^{\rm d} < 0$ or $F^{\rm d} = 0$ and $\dot{\lambda}^{\rm d} = 0$	$\mathbf{C}^{\rm epd} = \mathbf{C}^{\rm ed} + \mathbf{C}^{\rm pd}$
yes	yes	$F^{\rm p} > 0$ $(\dot{\lambda}^{\rm P} > 0)$	$F^{\rm d} > 0$ $(\dot{\lambda}^{\rm d} > 0)$	

Note that, in the above, the phenomenon of microcrack deactivation, when stress or strain changes sign from tensile to compressive, is not active.

2.2.4 Unilateral Damage Concepts

The procedure becomes much more complicated if *unilateral damage*, also called *damage deactivation* or *microcrack opening/closure*, are allowed for.

In the above it is assumed that the current state of damage at the beginning of a consecutive loading step is identical to that at the end of the preceding step $D_j^+ = D_{j+1}^-$, regardless of whether the stress (or strain) at the point considered changes from tensile to compressive, or not. It may happen, however, that when the material is (locally) subjected to tension–compression cycles, its initial stiffness is recovered, fully or partially, as shown schematically in Fig. 2.4.

In other words, the damage state remains "frozen" as long as, at the consecutive loading cycle, stress at the point considered again changes sign from compressive to tensile.

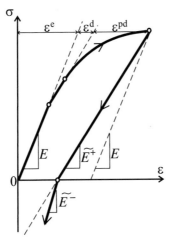

Fig. 2.4. Elastic stiffness recovery due to stress–controlled damage deactivation on changing stress sign

In order to properly describe the *unilateral damage conditions* a decomposition of the stress or strain tensors into positive and negative projections is usually introduced. To this end the fourth-rank *projection operators*, expressed in terms of principal directions of the strain or stress tensors, are used (Lubarda *et al.* [153], Hansen and Schreyer [95], Zhu and Cescotto [245], Krajcinovic [134])

$$\boldsymbol{\varepsilon}^+ = \mathbf{P}_\varepsilon^+ : \boldsymbol{\varepsilon}, \quad \boldsymbol{\varepsilon}^- = \mathbf{P}_\varepsilon^- : \boldsymbol{\varepsilon} = \left(\mathbf{I} - \mathbf{P}_\varepsilon^+\right) : \boldsymbol{\varepsilon} \qquad (2.49)$$

or

$$\boldsymbol{\sigma}^+ = \mathbf{P}_\sigma^+ : \boldsymbol{\sigma}, \quad \boldsymbol{\sigma}^- = \mathbf{P}_\sigma^- : \boldsymbol{\sigma} = \left(\mathbf{I} - \mathbf{P}_\sigma^+\right) : \boldsymbol{\sigma} \qquad (2.50)$$

where

$$\mathbf{P}_\varepsilon^+ = \sum_{i=1}^{3} \langle\langle\varepsilon^{(i)}\rangle\rangle \mathbf{n}_\varepsilon^{(i)} \otimes \mathbf{n}_\varepsilon^{(i)} \otimes \mathbf{n}_\varepsilon^{(i)} \otimes \mathbf{n}_\varepsilon^{(i)} \qquad (2.51)$$

or

$$\mathbf{P}_\sigma^+ = \sum_{i=1}^{3} \langle\langle\sigma^{(i)}\rangle\rangle \mathbf{n}_\sigma^{(i)} \otimes \mathbf{n}_\sigma^{(i)} \otimes \mathbf{n}_\sigma^{(i)} \otimes \mathbf{n}_\sigma^{(i)} \qquad (2.52)$$

The double angular bracket is defined as $\langle\langle a\rangle\rangle = 1$ for $a \geq 0$, or $\langle\langle a\rangle\rangle = 0$ for $a < 0$, and $\varepsilon^{(i)}$, $\sigma^{(i)}$ denote the principal strain or stress components. In other words, when the projection operators $\mathbf{P}_\varepsilon^{+/-}$ or $\mathbf{P}_\sigma^{+/-}$ are used, only the positive or negative principal components of the strain or stress tensors are extracted. When the simple unilateral damage condition is used it is assumed that negative principal components of the strain or stress tensors remain inactive as long as the loading conditions again render them active. A more general description of the unilateral damage condition allows for the influence of both positive and negative components of strain or stress tensors for damage evolution by using the additional *unilateral damage parameter* ζ_ε or ζ_σ (Murakami and Kamiya [162], Hayakawa and Murakami [98], Bielski *et al.* [22]). This concept reduces to introduction of the *generalized projection operators* (Ganczarski *et al.* [82])

$$\mathbf{P}_\varepsilon^* = \mathbf{P}_\varepsilon^+ + \zeta_\varepsilon \mathbf{P}_\varepsilon^- \qquad \text{or} \qquad \mathbf{P}_\sigma^* = \mathbf{P}_\sigma^+ + \zeta_\sigma \mathbf{P}_\sigma^- \qquad (2.53)$$

and the modified strain or stress tensors

$$\boldsymbol{\varepsilon}^* = \mathbf{P}_\varepsilon^* : \boldsymbol{\varepsilon} \qquad \text{or} \qquad \boldsymbol{\sigma}^* = \mathbf{P}_\sigma^* : \boldsymbol{\sigma} \qquad (2.54)$$

Note that when ζ_ε or ζ_σ are equal to 1, then $\mathbf{P}_\varepsilon^* = \mathbf{I}$ or $\mathbf{P}_\sigma^* = \mathbf{I}$, such that the unique mapping holds: $\boldsymbol{\varepsilon}^* = \boldsymbol{\varepsilon}$ or $\boldsymbol{\sigma}^* = \boldsymbol{\sigma}$ (no unilateral damage effect). In contrast, if ζ_ε or ζ_σ are equal to zero, then $\mathbf{P}_\varepsilon^* = \mathbf{P}_\varepsilon^+$ or $\mathbf{P}_\sigma^* = \mathbf{P}_\sigma^+$, hence negative principal components of strain or stress tensors have no influence on damage evolution, such that $\boldsymbol{\varepsilon}^* = \boldsymbol{\varepsilon}^+$ or $\boldsymbol{\sigma}^* = \boldsymbol{\sigma}^+$ hold (Litewka [150]). When general coordinate systems are used, the modified strain or stress tensors are expressed in terms of the actual ones by the following mappings (Bielski *et al.* [22])

$$\varepsilon_{ij}^* = B_{ijkl}^{(\varepsilon)} \varepsilon_{kl} \qquad B_{ijkl}^{(\varepsilon)} = \sum_{I=1}^3 \xi(\varepsilon_I) n_{iI}^{(\varepsilon)} n_{jI}^{(\varepsilon)} n_{Ik}^{(\varepsilon)} n_{Il}^{(\varepsilon)} \qquad (2.55)$$

or

$$\sigma_{ij}^* = B_{ijkl}^{(\sigma)} \sigma_{kl} \qquad B_{ijkl}^{(\sigma)} = \sum_{I=1}^3 \xi(\sigma_I) n_{iI}^{(\sigma)} n_{jI}^{(\sigma)} n_{Ik}^{(\sigma)} n_{Il}^{(\sigma)} \qquad (2.56)$$

where the fourth-rank tensors $B_{ijkl}^{(\varepsilon)}$ or $B_{ijkl}^{(\sigma)}$ are built of direction cosines between the principal and the current spatial systems.

Limitations of the unilateral damage condition are discussed by Chaboche [34, 35], Chaboche *et al.* [40], Halm and Dragon [93]. Some existing concepts (Krajcinovic and Fonseka [135], Ju [121]) may lead to nonsymmetric effective elastic stiffness or compliance, or to stress–strain discontinuities under non-proportional loading. It was shown in Skrzypek and Kuna-Ciskał [207] that, if the unilateral damage condition influences both diagonal and off-diagonal components of the stiffness or compliance matrix, a stress discontinuity occurs when one of the principal strains changes sign, whereas the others remain unchanged. It means, that the above defined modified strain or stress tensors (2.55) or (2.56) can be built-in to the state potential Ψ or Γ in such a way that only diagonal components of the constitutive matrices are affected by unilateral damage, e.g. $[\mathbf{C}^*(\boldsymbol{\sigma}^*, \mathbf{D}, \mathbf{Y}, r, T)]_{ii}$

$$\begin{Bmatrix} \dot{\varepsilon}_{11} - \widetilde{\alpha}_{11}\dot{T} \\ \dot{\varepsilon}_{22} - \widetilde{\alpha}_{22}\dot{T} \\ \dot{\varepsilon}_{33} - \widetilde{\alpha}_{33}\dot{T} \\ \dot{\gamma}_{23} \\ \dot{\gamma}_{13} \\ \dot{\gamma}_{12} \end{Bmatrix} = \begin{bmatrix} C_{11}^*(\zeta) & C_{12} & C_{13} & 0 & 0 & 0 \\ & C_{22}^*(\zeta) & C_{23} & 0 & 0 & 0 \\ & & C_{33}^*(\zeta) & 0 & 0 & 0 \\ & & & C_{44}^*(\zeta) & 0 & 0 \\ & & & & C_{55}^*(\zeta) & 0 \\ & & & & & C_{66}^*(\zeta) \end{bmatrix} \begin{Bmatrix} \dot{\sigma}_{11} \\ \dot{\sigma}_{22} \\ \dot{\sigma}_{33} \\ \dot{\sigma}_{23} \\ \dot{\sigma}_{13} \\ \dot{\sigma}_{12} \end{Bmatrix}$$

$$(2.57)$$

By contrast, the off-diagonal components of the constitutive matrix must be free from the damage deactivation effect in order to properly describe the unilateral damage condition in such a way that stress continuity holds.

2.2.5 Continuous Damage Deactivation

Uniaxial case

Inconsistencies of the unilateral damage theory discussed in the previous section may be successfully eliminated by application of *continuous damage deactivation*. The fundamentals of this concept, based on work by Lemaitre [147] and Hansen and Hansen and Schreyer [95] is the subject of a series of papers by Foryś and Ganczarski [81], Ganczarski and Barwacz [83] and Ganczarski and Cegielski [84].

In the case of uniaxial tensile stress and scalar damage, the effective stress and the appropriate effective modulus of elasticity are defined as follows:

$$\tilde{\sigma} = \sigma/(1-D) \qquad \tilde{E} = E(1-D) \tag{2.58}$$

The above relations are valid also in cases when microcracks remain open under uniaxial compression. For a certain class of materials and certain conditions of loading the microdefects may close in compression. This is often the case for very brittle materials. If the microdefects close completely two separate sets of conditions must be defined, one for tension and another for compression:

$$\tilde{\sigma}^{\pm} = \begin{cases} \sigma/(1-D) \\ \sigma \end{cases} \qquad \tilde{E}^{\pm} = \begin{cases} E(1-D) & \text{tension} \\ E & \text{compression} \end{cases} \tag{2.59}$$

In the real material, however, the microdefects have complicated shapes and do not close completely. In order to take into account this effect, the so called crack closure parameter h $(0 \leq h \leq 1)$ is introduced. The *crack closure parameter* h depends on material and loading, however, in practice it is considered to be constant $h = h_c$ and its universal value applicable for the majority of engineering materials is equal to $h_c = 0.2$ (Lemaitre [147]). Therefore appropriate conditions for tension and compression are given by the following formulae

$$\tilde{\sigma}^{\pm} = \begin{cases} \sigma/(1-Dh) \\ \sigma/(1-Dh_c) \end{cases} \qquad \tilde{E}^{\pm} = \begin{cases} E(1-Dh) & \text{tension} \\ E(1-Dh_c) & \text{compression} \end{cases} \tag{2.60}$$

Application of this model for the description of the unloading path leads to a linear relation between the stress decrease and the strain decrease given by $\tilde{E}^+$. Entering the compression range the material switches to the path characterized by the modulus of elasticity which is equal to $\tilde{E}^-$ (Fig. 2.5a). The real materials do not exhibit such bilinear unloading paths. The concept of *continuous crack closure* proposed by Hansen and Schreyer [95], which allows

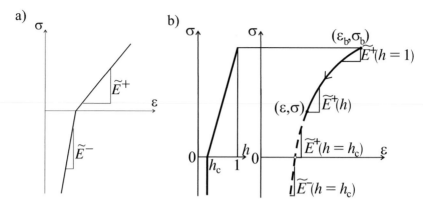

Fig. 2.5. Bilinear unloading path: (a) effect of continuous damage deactivation; (b) (Ganczarski and Barwacz [83] and Ganczarski and Cegielski [84])

one to eliminate the switch between $\widetilde{E}^+$ and $\widetilde{E}^-$, and also to consider the effect of hardening accompanying the crack closure, consists in the replacement of parameter h by a function $h(\sigma)$, linear in the simplest case, such that

$$h(\sigma) = h_{\mathrm{c}} + (1 - h_{\mathrm{c}}) \frac{\langle \sigma \rangle}{\sigma_{\mathrm{b}}} \tag{2.61}$$

where σ_{b} denotes stress referred to the beginning of unloading. Corresponding *modified modulus of elasticity* for the unloading path is defined as follows:

$$\widetilde{E}_{\mathrm{u}} = E \left\{ 1 - D \left[h_{\mathrm{c}} + (1 - h_{\mathrm{c}}) \frac{\langle \sigma \rangle}{\sigma_{\mathrm{b}}} \right] \right\} \tag{2.62}$$

Application of the modulus defined by Eq. (2.62) to the modelling of the unloading path leads to the hyperbolic-type relation between strain and stress (Fig. 2.5b).

General 3D stress state

Distinction between tension and compression, being the key point of the general 3D extension, is not trivial. However, when the stress tensor is given by its eigenvalues, the following decomposition is applied:

$$\boldsymbol{\sigma} = \begin{pmatrix} \sigma_1 & 0 & 0 \\ 0 & \sigma_2 & 0 \\ 0 & 0 & \sigma_3 \end{pmatrix} = \langle \boldsymbol{\sigma} \rangle - \langle -\boldsymbol{\sigma} \rangle \tag{2.63}$$

which, in the case of isotropic damage and application of the *strain equivalence principle*, results in the following formula for the effective stress:

$$\widetilde{\boldsymbol{\sigma}}^{+} = \frac{\langle\boldsymbol{\sigma}\rangle}{1 - Dh\left(\boldsymbol{\sigma}\right)} + \frac{\nu}{1 - 2\nu} \frac{\left[\mathrm{Tr}\left\langle\boldsymbol{\sigma}\right\rangle - \left\langle\mathrm{Tr}\left(\boldsymbol{\sigma}\right)\right\rangle\right]}{1 - Dh\left(\boldsymbol{\sigma}\right)} \mathbf{1}$$

$$\widetilde{\boldsymbol{\sigma}}^{-} = -\frac{\langle-\boldsymbol{\sigma}\rangle}{1 - Dh_{\mathrm{c}}} - \frac{\nu}{1 - 2\nu} \frac{\left[\mathrm{Tr}\left\langle-\boldsymbol{\sigma}\right\rangle - \left\langle-\mathrm{Tr}\left(\boldsymbol{\sigma}\right)\right\rangle\right]}{1 - Dh_{\mathrm{c}}} \mathbf{1}$$

(2.64)

Terms associated to the factor $\frac{\nu}{1-2\nu}$ introducing coupling disappear if all eigenvalues of stress are of the same sign

$$\begin{array}{llllll} \sigma_1 > 0 & \wedge & \sigma_2 > 0 & \wedge & \sigma_3 > 0 & \Rightarrow & \mathrm{Tr}\left\langle\boldsymbol{\sigma}^{+}\right\rangle = \left\langle\mathrm{Tr}\left(\boldsymbol{\sigma}\right)\right\rangle \\ \sigma_1 < 0 & \wedge & \sigma_2 < 0 & \wedge & \sigma_3 < 0 & \Rightarrow & \mathrm{Tr}\left\langle-\boldsymbol{\sigma}^{-}\right\rangle = \left\langle-\mathrm{Tr}\left(\boldsymbol{\sigma}\right)\right\rangle \end{array}$$

(2.65)

In such a case, simplified effective stresses and the corresponding effective elastic tensors take the form

$$\widetilde{\boldsymbol{\sigma}}^{\pm} = \begin{cases} \left\langle\boldsymbol{\sigma}^{+}\right\rangle / \left(1 - Dh\left(\boldsymbol{\sigma}\right)\right) \\ -\left\langle-\boldsymbol{\sigma}^{-}\right\rangle / \left(1 - Dh_{\mathrm{c}}\right) \end{cases} \quad \widetilde{\mathbf{E}}^{\pm} = \begin{cases} \mathbf{E}\left(1 - Dh\left(\boldsymbol{\sigma}\right)\right) & \text{tension} \\ \mathbf{E}\left(1 - Dh_{\mathrm{c}}\right) & \text{compression} \end{cases}$$

(2.66)

where $\mathbf{E}$ is the Hooke tensor (matrix)

$$[\mathbf{E}] = \frac{2G}{1 - 2\nu} \begin{bmatrix} 1 - \nu & \nu & \nu & & & \\ \nu & 1 - \nu & \nu & & & \\ \nu & \nu & 1 - \nu & & & \\ & & & 1 - 2\nu & & \\ & & & & 1 - 2\nu & \\ & & & & & 1 - 2\nu \end{bmatrix}$$

(2.67)

Application of the concept of continuous crack closure needs an additional hypothesis that introduces functional coupling between the crack closure parameter and a scalar function of the stress tensor $\chi(\boldsymbol{\sigma})$:

$$h\left(\boldsymbol{\sigma}\right) = h_{\mathrm{c}} + \left(1 - h_{\mathrm{c}}\right) \frac{\chi\left(\boldsymbol{\sigma}\right)}{\chi\left(\boldsymbol{\sigma}_{\mathrm{b}}\right)}$$

(2.68)

where $\chi\left(\boldsymbol{\sigma}\right)$ denotes the known Hayhurst function

$$\chi\left(\boldsymbol{\sigma}\right) = \alpha\mathrm{Tr}\left(\boldsymbol{\sigma}\right) + \left(1 - \alpha\right)J_2\left(\boldsymbol{\sigma}\right)$$

(2.69)

and again $\boldsymbol{\sigma}_{\mathrm{b}}$ denotes stress referred to the beginning of unloading.

If damage is described by the *second-rank damage tensor* represented by its eigenvalues Murakami and Ohno [163]

$$\mathbf{D} = \begin{pmatrix} D_1 & 0 & 0 \\ 0 & D_2 & 0 \\ 0 & 0 & D_3 \end{pmatrix}$$

(2.70)

and when the eigendirections of damage coincide with the eigendirections of stress, the *effective stresses extension* of (2.64) is as follows

$$\tilde{\sigma}^+ = \left\langle \boldsymbol{\sigma} \cdot (1 - \mathbf{D} \cdot \mathbf{h}(\boldsymbol{\sigma}))^{-1} \right\rangle$$
$$+ \frac{\nu}{1 - 2\nu} \left\{ \mathrm{Tr} \left\langle \boldsymbol{\sigma} \cdot (1 - \mathbf{D} \cdot \mathbf{h}(\boldsymbol{\sigma}))^{-1} \right\rangle - \left\langle \mathrm{Tr} \left[\boldsymbol{\sigma} \cdot (1 - \mathbf{D} \cdot \mathbf{h}(\boldsymbol{\sigma}))^{-1} \right] \right\rangle \right\}$$
$$\tilde{\sigma}^- = - \left\langle -\boldsymbol{\sigma} \cdot (1 - \mathbf{D} \cdot \mathbf{h}_c)^{-1} \right\rangle$$
$$- \frac{\nu}{1 - 2\nu} \left\{ \mathrm{Tr} \left\langle -\boldsymbol{\sigma} \cdot (1 - \mathbf{D} \cdot \mathbf{h}_c)^{-1} \right\rangle - \left\langle -\mathrm{Tr} \left[\boldsymbol{\sigma} \cdot (1 - \mathbf{D} \cdot \mathbf{h}_c)^{-1} \right] \right\rangle \right\}$$

$$(2.71)$$

where $\mathbf{h}(\boldsymbol{\sigma})$ stands for the second-rank crack closure tensor

$$\mathbf{h}(\boldsymbol{\sigma}) = \begin{pmatrix} h_1(\boldsymbol{\sigma}) & 0 & 0 \\ 0 & h_2(\boldsymbol{\sigma}) & 0 \\ 0 & 0 & h_3(\boldsymbol{\sigma}) \end{pmatrix} \qquad (2.72)$$

and tensor $\mathbf{h}_c$ is equal to

$$\mathbf{h}_c = \begin{pmatrix} h_c & 0 & 0 \\ 0 & h_c & 0 \\ 0 & 0 & h_c \end{pmatrix} \qquad (2.73)$$

Again, if all eigenvalues of the stress are of the same sign, the terms with factor $\frac{\nu}{1-2\nu}$ disappear, such that (2.71) takes the simpler form

$$\tilde{\sigma}^\pm = \begin{cases} \left\langle \boldsymbol{\sigma} \cdot (1 - \mathbf{D} \cdot \mathbf{h}(\boldsymbol{\sigma}))^{-1} \right\rangle & \text{tension} \\ -\left\langle -\boldsymbol{\sigma} \cdot (1 - \mathbf{D} \cdot \mathbf{h}_c)^{-1} \right\rangle & \text{compression} \end{cases} \qquad (2.74)$$

Corresponding modified elastic tensors, including the continuous crack closure effect, are given by the formulae

$$\tilde{\mathbf{E}}^\pm = \frac{1}{2} \left[\mathbf{E}{:}\mathbf{M}^\pm(\mathbf{D}) + \mathbf{M}^\pm(\mathbf{D}){:}\mathbf{E} \right] \qquad (2.75)$$

that take advantage of the fourth-rank damage effect tensor defined as

$$[\mathbf{M}^\pm(\mathbf{D})] =$$
$$\begin{bmatrix} 1 - M_1 & 0 & 0 & 0 & 0 & 0 \\ 0 & 1 - M_2 & 0 & 0 & 0 & 0 \\ 0 & 0 & 1 - M_1 & 0 & 0 & 0 \\ 0 & 0 & 0 & 1 - \frac{M_2 + M_3}{2} & 0 & 0 \\ 0 & 0 & 0 & 0 & 1 - \frac{M_1 + M_3}{2} & 0 \\ 0 & 0 & 0 & 0 & 0 & 1 - \frac{M_1 + M_2}{2} \end{bmatrix} \qquad (2.76)$$

where $M_i = D_i(1 - h_c) \frac{\langle \sigma_i \rangle}{\sigma_{i_b}}$.

3

Developing and Implementing Selected Constitutive Models for Elasto-Plastic-Damage Materials

3.1 The Constitutive Model of Elastic-Damage Material

3.1.1 Incremental Formulation of the Constitutive Equations

Basic equations

The general thermodynamically based theory of the constitutive and evolution equations of elastic–brittle damaged materials in a total stress–strain formulation, developed by Murakami and Kamiya [162], is the key for a further extension presented in the paper by Kuna-Ciskał and Skrzypek [208]. The *Helmholtz free energy* is a function of the elastic strain tensor $\boldsymbol{\varepsilon}^e$, the second-rank damage tensor $\mathbf{D}$, and another scalar damage variable β. The following Helmholtz free energy decomposition into the elastic and the damage terms is postulated as the *state potential* (2.20):

$$\rho\Psi(\boldsymbol{\varepsilon}^e, \mathbf{D}, \beta) = \rho\Psi^e(\boldsymbol{\varepsilon}^e, \mathbf{D}) + \rho\Psi^d(\beta) \tag{3.1}$$

In order to properly describe the unilateral damage response in tension or compression a *modified elastic strain tensor* $\boldsymbol{\varepsilon}^{*e}$ in the principal strain coordinate system is defined

$$\begin{aligned} \varepsilon_I^{*e} &= \langle \varepsilon_I^e \rangle - \zeta \langle -\varepsilon_I^e \rangle = k(\varepsilon_I^e)\varepsilon_I^e & \zeta \in \langle 0, 1 \rangle \\ k(\varepsilon_I^e) &= k_I = H(\varepsilon_I^e) + \zeta H(-\varepsilon_I^e) & I = 1, 2, 3 \end{aligned} \tag{3.2}$$

In other words, the modified elastic strain tensor in the principal directions is given by

$$\boldsymbol{\varepsilon}^{*e} = \left\{ \begin{array}{ccc} k_1\varepsilon_1^e & & \\ & k_2\varepsilon_2^e & \\ & & k_3\varepsilon_3^e \end{array} \right\} \tag{3.3}$$

where coefficients k_1, k_2 and k_3 are equal to 1 or ζ depending on whether the corresponding strain takes the positive or negative principal value, respectively.

Symbol $\langle\rangle$ denotes the *Macauley bracket*, $H()$ is the Heaviside step function, $\varepsilon_{\mathrm{I}}(\mathrm{I}=1,2,3)$ are principal values of $\boldsymbol{\varepsilon}^{\mathrm{e}}$, and ζ is an additional material constant responsible for the unilateral damage response effect under tension or compression (Krajcinovic [133]). For $\zeta=1$ the modified strain tensor $\boldsymbol{\varepsilon}^{*\mathrm{e}}$ is identical to $\boldsymbol{\varepsilon}^{\mathrm{e}}$ and the unilateral damage (crack opening/closure) effect is not accounted for. In contrast, for $\zeta=0$, the strain tensor $\boldsymbol{\varepsilon}^{*\mathrm{e}}$ is modified in such a way that negative principal strain components are replaced by zeros, whereas positive ones remain unchanged.

Following the Murakami and Kamiya [162] assumptions, both terms of the free energy (3.1) are represented as

$$\rho\Psi^{\mathrm{e}}(\boldsymbol{\varepsilon}^{\mathrm{e}},\mathbf{D})=\frac{1}{2}\lambda(\mathrm{Tr}\varepsilon^{\mathrm{e}})^2+\mu\mathrm{Tr}(\boldsymbol{\varepsilon}^{\mathrm{e}}\cdot\boldsymbol{\varepsilon}^{\mathrm{e}})+\eta_1(\mathrm{Tr}\varepsilon^{\mathrm{e}})^2\mathrm{Tr}\mathbf{D}$$

$$+\eta_2\mathrm{Tr}(\boldsymbol{\varepsilon}^{\mathrm{e}}\cdot\boldsymbol{\varepsilon}^{\mathrm{e}})\mathrm{Tr}\mathbf{D}+\eta_3\mathrm{Tr}\varepsilon^{\mathrm{e}}\mathrm{Tr}(\boldsymbol{\varepsilon}^{\mathrm{e}}\cdot\mathbf{D})+\eta_4\mathrm{Tr}\left[\boldsymbol{\varepsilon}^{*\mathrm{e}}\cdot\boldsymbol{\varepsilon}^{*\mathrm{e}}\cdot\mathbf{D}\right] \quad (3.4)$$

$$\rho\Psi^{\mathrm{d}}(\beta)=\tfrac{1}{2}K_{\mathrm{d}}\beta^2$$

where λ and μ are Lamè constants for undamaged materials, $\eta_1,\eta_2,\eta_3,\eta_4$ and K_{d} are material constants. Note that the modified strain tensor $\boldsymbol{\varepsilon}^{*\mathrm{e}}$ is applied in the last term of Ψ^{e} exclusively, which ensures the continuous transition of the stress–strain response from crack opening to closure (Chaboche [35], Chaboche *et al.* [40]).

The following constitutive equation of anisotropic elasticity coupled with damage is obtained from (3.4) according to the conventional procedure of thermodynamic formalism, described in Sect. 2.2.3:

$$\boldsymbol{\sigma}=\frac{\partial(\rho\Psi)}{\partial\boldsymbol{\varepsilon}^{\mathrm{e}}}=\left[\lambda(\mathrm{Tr}\varepsilon^{\mathrm{e}})+2\eta_1\mathrm{Tr}\varepsilon^{\mathrm{e}}\mathrm{Tr}\mathbf{D}+\eta_3\mathrm{Tr}(\boldsymbol{\varepsilon}^{\mathrm{e}}\cdot\mathbf{D})\right]\mathbf{1}$$

$$+2\left[\mu+\eta_2\mathrm{Tr}\mathbf{D}\right]\boldsymbol{\varepsilon}^{\mathrm{e}}+\eta_3\mathrm{Tr}(\boldsymbol{\varepsilon}^{\mathrm{e}})\mathbf{D}+\eta_4\left(\boldsymbol{\varepsilon}^{*\mathrm{e}}\cdot\mathbf{D}+\mathbf{D}\cdot\boldsymbol{\varepsilon}^{*\mathrm{e}}\right):\frac{\partial\boldsymbol{\varepsilon}^{*\mathrm{e}}}{\partial\boldsymbol{\varepsilon}^{\mathrm{e}}} \quad (3.5)$$

$$=^{\mathrm{s}}\widetilde{\mathbf{S}}(\mathbf{D}):\boldsymbol{\varepsilon}^{\mathrm{e}}$$

The thermodynamic-force conjugates of $\mathbf{D}$ and β are also derived from (3.4):

$$\mathbf{Y}=-\rho\frac{\partial\Psi^{\mathrm{e}}}{\partial\mathbf{D}}=-\left[\eta_1(\mathrm{Tr}\varepsilon^{\mathrm{e}})^2+\eta_2\mathrm{Tr}(\boldsymbol{\varepsilon}^{\mathrm{e}}\cdot\boldsymbol{\varepsilon}^{\mathrm{e}})\right]\mathbf{1}-\eta_3(\mathrm{Tr}\varepsilon^{\mathrm{e}})\boldsymbol{\varepsilon}^{\mathrm{e}}-\eta_4\boldsymbol{\varepsilon}^{*\mathrm{e}}\cdot\boldsymbol{\varepsilon}^{*\mathrm{e}}$$

$$B=\rho\frac{\partial\Psi^{\mathrm{d}}}{\partial\beta}=K_{\mathrm{d}}\beta$$

$$(3.6)$$

The damage dissipation potential in the space of force conjugates $\{\mathbf{Y},-B\}$ is assumed in the form

$$F^{\mathrm{d}}(\mathbf{Y},B)=Y_{\mathrm{eq}}-(B_0+B)=0$$
$$Y_{\mathrm{eq}}=\sqrt{\tfrac{1}{2}\mathbf{Y}:\mathbf{L}:\mathbf{Y}}$$
$$L_{ijkl}=\tfrac{1}{2}(\delta_{ik}\delta_{jl}+\delta_{il}\delta_{jk})$$

$$(3.7)$$

where B_0 and B denote the initial damage threshold and the subsequent damage force conjugates of β, respectively, and a reduced form of (2.29) is used. Hence, experimentally motivated by Murakami and Kamiya [162], the isotropic hardening of damage dissipation surface (3.7) in the $\{Y, -B\}$ space is introduced.

The damage evolution equations are finally established from the normality rule:

$$\dot{\mathbf{D}} = \dot{\lambda}^{\mathrm{d}} \frac{\partial F^{\mathrm{d}}}{\partial \mathbf{Y}}, \qquad \dot{\beta} = \dot{\lambda}^{\mathrm{d}} \frac{\partial F^{\mathrm{d}}}{\partial(-B)} = \dot{\lambda}^{\mathrm{d}} \tag{3.8}$$

and the consistency condition $\dot{F}^{\mathrm{d}} = 0 = \dfrac{\partial F^{\mathrm{d}}}{\partial \mathbf{Y}} : \dot{\mathbf{Y}} + \dfrac{\partial F^{\mathrm{d}}}{\partial B} \dot{B}$ is used to eliminate $\dot{\lambda}^{\mathrm{d}}$ from (3.8):

$$\dot{\lambda}^{\mathrm{d}} = \frac{\alpha \dfrac{\partial F^{\mathrm{d}}}{\partial \mathbf{Y}} : \dot{\mathbf{Y}}}{\dfrac{\partial B}{\partial \beta}} = \alpha \frac{\mathbf{L} : \mathbf{Y}}{2 K_{\mathrm{d}} Y_{\mathrm{eq}}} : \dot{\mathbf{Y}} \tag{3.9}$$

A factor $\alpha = 1$ or $\alpha = 0$ is used for the active or passive damage growth, respectively.

Incremental formulation of the modified elastic-damage model

When the total formulation is used the constitutive equation (3.5) represents stress–strain relations by the use of the *secant elastic damage stiffness* $^{\mathrm{s}}\widetilde{\mathbf{S}}(\mathbf{D})$, which suffers from damage evolution in a material (2.33)

$$\boldsymbol{\sigma} = {}^{\mathrm{s}} \widetilde{\mathbf{S}} : \boldsymbol{\varepsilon}^{\mathrm{e}} \qquad \text{or} \qquad \sigma_{ij} = {}^{\mathrm{s}} \widetilde{S}_{ijkl} : \varepsilon^{\mathrm{e}}_{kl} \tag{3.10}$$

The incremental form of the stress–strain relations may be established from (3.10) as follows:

$$\mathrm{d}\boldsymbol{\sigma} = {}^{\mathrm{s}} \widetilde{\mathbf{S}} : \mathrm{d}\boldsymbol{\varepsilon}^{\mathrm{e}} + \boldsymbol{\varepsilon}^{\mathrm{e}} : \left(\frac{\partial {}^{\mathrm{s}}\widetilde{\mathbf{S}}}{\partial \mathbf{D}} : \mathrm{d}\mathbf{D} \right) \tag{3.11}$$

The damage increment $\mathrm{d}\mathbf{D}$ on $\mathrm{d}\boldsymbol{\varepsilon}^{\mathrm{e}}$ is obtained from (3.8)–(3.9). Eventually, the following *incremental state equation* is derived (Skrzypek and Kuna-Ciskał [207, 208]):

$$\mathrm{d}\boldsymbol{\sigma} = \left[{}^{\mathrm{s}}\widetilde{\mathbf{S}} + \alpha \boldsymbol{\varepsilon}^{\mathrm{e}} : \frac{\partial {}^{\mathrm{s}}\widetilde{\mathbf{S}}}{\partial \mathbf{D}} : \left(\frac{\mathbf{Y}\mathbf{Y} : \dfrac{\partial \mathbf{Y}}{\partial \boldsymbol{\varepsilon}^{\mathrm{e}}}}{4 K_{\mathrm{d}} Y_{\mathrm{eq}}^2} \right) \right] : \mathrm{d}\boldsymbol{\varepsilon}^{\mathrm{e}} \tag{3.12}$$

The square bracket in (3.12) represents the *effective tangent stiffness* $\widetilde{\mathbf{S}}(\boldsymbol{\varepsilon}^{\mathrm{e}}, \mathbf{D})$ and factor α equal to 0 or 1 is used for a passive or active damage process, respectively. By the use of *Voigt notation* the state equations, total (3.10) and incremental (3.12), may easily be rewritten in more convenient matrix forms:

$$\{\boldsymbol{\sigma}\} = \left[{}^{\mathrm{s}}\widetilde{\mathbf{S}}(\mathbf{D})\right]\{\varepsilon^{\mathrm{e}}\} \qquad \text{or} \qquad \sigma_i = {}^{\mathrm{s}}\widetilde{S}_{ij}\varepsilon_j^{\mathrm{e}}$$
$$\{\mathrm{d}\boldsymbol{\sigma}\} = \left[\widetilde{\mathbf{S}}(\varepsilon^{\mathrm{e}},\mathbf{D})\right]\{\mathrm{d}\varepsilon^{\mathrm{e}}\} \qquad \text{or} \qquad \mathrm{d}\sigma_i = \widetilde{S}_{ij}\mathrm{d}\varepsilon_j^{\mathrm{e}} \tag{3.13}$$

In a general 3D case the following matrix representation of the *total constitutive equation* holds:

$$\begin{Bmatrix} \sigma_{11} \\ \sigma_{22} \\ \sigma_{33} \\ \sigma_{23} \\ \sigma_{13} \\ \sigma_{12} \end{Bmatrix} = \begin{bmatrix} {}^{\mathrm{s}}\widetilde{S}_{11} & {}^{\mathrm{s}}\widetilde{S}_{12} & {}^{\mathrm{s}}\widetilde{S}_{13} & {}^{\mathrm{s}}\widetilde{S}_{14} & {}^{\mathrm{s}}\widetilde{S}_{15} & {}^{\mathrm{s}}\widetilde{S}_{16} \\ & {}^{\mathrm{s}}\widetilde{S}_{22} & {}^{\mathrm{s}}\widetilde{S}_{23} & {}^{\mathrm{s}}\widetilde{S}_{24} & {}^{\mathrm{s}}\widetilde{S}_{25} & {}^{\mathrm{s}}\widetilde{S}_{26} \\ & & {}^{\mathrm{s}}\widetilde{S}_{33} & {}^{\mathrm{s}}\widetilde{S}_{34} & {}^{\mathrm{s}}\widetilde{S}_{35} & {}^{\mathrm{s}}\widetilde{S}_{36} \\ & & & {}^{\mathrm{s}}\widetilde{S}_{44} & {}^{\mathrm{s}}\widetilde{S}_{45} & {}^{\mathrm{s}}\widetilde{S}_{46} \\ & \text{symm.} & & & {}^{\mathrm{s}}\widetilde{S}_{55} & {}^{\mathrm{s}}\widetilde{S}_{56} \\ & & & & & {}^{\mathrm{s}}\widetilde{S}_{66} \end{bmatrix} \begin{Bmatrix} \varepsilon_{11}^{\mathrm{e}} \\ \varepsilon_{22}^{\mathrm{e}} \\ \varepsilon_{33}^{\mathrm{e}} \\ \gamma_{23}^{\mathrm{e}} \\ \gamma_{13}^{\mathrm{e}} \\ \gamma_{12}^{\mathrm{e}} \end{Bmatrix} \tag{3.14}$$

The "full" 6×6 symmetric *secant stiffness matrix* is as follows:

$$\begin{aligned}
{}^{\mathrm{s}}\widetilde{S}_{11} &= \lambda + 2\mu + 2(\eta_1 + \eta_2)\mathrm{Tr}\mathbf{D} + 2(\eta_3 + \eta_4)D_{11} \\
{}^{\mathrm{s}}\widetilde{S}_{22} &= \lambda + 2\mu + 2(\eta_1 + \eta_2)\mathrm{Tr}\mathbf{D} + 2(\eta_3 + \eta_4)D_{22} \\
{}^{\mathrm{s}}\widetilde{S}_{33} &= \lambda + 2\mu + 2(\eta_1 + \eta_2)\mathrm{Tr}\mathbf{D} + 2(\eta_3 + \eta_4)D_{33} \\
{}^{\mathrm{s}}\widetilde{S}_{12} &= \lambda + 2\eta_1\mathrm{Tr}\mathbf{D} + \eta_3(D_{11} + D_{22}) \\
{}^{\mathrm{s}}\widetilde{S}_{13} &= \lambda + 2\eta_1\mathrm{Tr}\mathbf{D} + \eta_3(D_{11} + D_{33}) \\
{}^{\mathrm{s}}\widetilde{S}_{23} &= \lambda + 2\eta_1\mathrm{Tr}\mathbf{D} + \eta_3(D_{22} + D_{33}) \\
{}^{\mathrm{s}}\widetilde{S}_{44} &= \tfrac{1}{2}\left[2\mu + 2\eta_2\mathrm{Tr}\mathbf{D} + \eta_4(D_{33} + D_{22})\right] \\
{}^{\mathrm{s}}\widetilde{S}_{45} &= \eta_4 D_{12} \\
{}^{\mathrm{s}}\widetilde{S}_{55} &= \tfrac{1}{2}\left[2\mu + 2\eta_2\mathrm{Tr}\mathbf{D} + \eta_4(D_{11} + D_{33})\right] \\
{}^{\mathrm{s}}\widetilde{S}_{46} &= \eta_4 D_{13} \\
{}^{\mathrm{s}}\widetilde{S}_{66} &= \tfrac{1}{2}\left[2\mu + 2\eta_2\mathrm{Tr}\mathbf{D} + \eta_4(D_{11} + D_{22})\right] \\
{}^{\mathrm{s}}\widetilde{S}_{56} &= \eta_4 D_{23} \\
{}^{\mathrm{s}}\widetilde{S}_{14} &= \eta_3 D_{23} \\
{}^{\mathrm{s}}\widetilde{S}_{24} &= {}^{\mathrm{s}}\widetilde{S}_{34} = (\eta_3 + \eta_4)D_{23} \\
{}^{\mathrm{s}}\widetilde{S}_{25} &= \eta_3 D_{13} \\
{}^{\mathrm{s}}\widetilde{S}_{15} &= {}^{\mathrm{s}}\widetilde{S}_{35} = (\eta_3 + \eta_4)D_{13} \\
{}^{\mathrm{s}}\widetilde{S}_{36} &= \eta_3 D_{12} \\
{}^{\mathrm{s}}\widetilde{S}_{16} &= {}^{\mathrm{s}}\widetilde{S}_{26} = (\eta_3 + \eta_4)D_{12}
\end{aligned} \tag{3.15}$$

Let us mention that the above "general" formulae are limited to a special case $\zeta = 1$, where $\varepsilon^{\mathrm{e}} \equiv \varepsilon^{*\mathrm{e}}$ (the unilateral damage effect is ignored in (3.15)). Representation of the effective tangent stiffness matrix in a general 3D state with unilateral effect included is rather cumbersome hence, in what follows, we confine ourselves to the simpler plane stress state (for full representation see Skrzypek and Kuna-Ciskał [207]).

Matrix constitutive equations in a plane stress state

When the assumption of *plane stress* is used $\sigma_{33} = 0$, the total form of the *matrix constitutive equations* (3.14) reduces to

$$\left\{\begin{array}{c} \sigma_{11} \\ \sigma_{22} \\ 0 \\ \sigma_{12} \end{array}\right\} = \left[\begin{array}{cccc} {}^s\widetilde{S}_{11} & {}^s\widetilde{S}_{12} & {}^s\widetilde{S}_{13} & {}^s\widetilde{S}_{16} \\ & {}^s\widetilde{S}_{22} & {}^s\widetilde{S}_{23} & {}^s\widetilde{S}_{26} \\ & & {}^s\widetilde{S}_{33} & {}^s\widetilde{S}_{36} \\ & & & {}^s\widetilde{S}_{66} \end{array}\right] \left\{\begin{array}{c} \varepsilon_{11}^e \\ \varepsilon_{22}^e \\ \varepsilon_{33}^e \\ \gamma_{12}^e \end{array}\right\} \tag{3.16}$$

where the symmetric 4×4 effective secant stiffness matrix, which depends on **D**, is as follows (Kuna-Ciskał and Skrzypek [208]):

$$\begin{aligned} {}^s\widetilde{S}_{11} &= \lambda + 2\mu + 2(\eta_1 + \eta_2)\mathrm{Tr}\mathbf{D} + 2\eta_3 D_{11} + \eta_4 \left[D_{11}\left(2a^2 + \tfrac{1}{2}b^2\right) \right. \\ &\left. + \tfrac{1}{2}D_{22}b^2 + 2abD_{12}\right] \end{aligned}$$

$$\begin{aligned} {}^s\widetilde{S}_{22} &= \lambda + 2\mu + 2(\eta_1 + \eta_2)\mathrm{Tr}\mathbf{D} + 2\eta_3 D_{22} + \eta_4 \left[\tfrac{1}{2}b^2 D_{11} \right. \\ &\left. + D_{22}\left(2c^2 + \tfrac{1}{2}b^2\right) + 2bcD_{12}\right] \end{aligned}$$

$$ {}^s\widetilde{S}_{33} = \lambda + 2\mu + 2(\eta_1 + \eta_2)\mathrm{Tr}\mathbf{D} + 2(\eta_3 + \eta_4 k_3^2)D_{33} $$

$$\begin{aligned} {}^s\widetilde{S}_{12} &= {}^s\widetilde{S}_{21} = \lambda + 2\eta_1 \mathrm{Tr}\mathbf{D} + \eta_3(D_{11} + D_{22}) \\ &+ \eta_4 \left[\tfrac{1}{2}b^2(D_{11} + D_{22}) + D_{12}b(a + c)\right] \end{aligned}$$

$$ {}^s\widetilde{S}_{13} = {}^s\widetilde{S}_{31} = \lambda + 2\eta_1 \mathrm{Tr}\mathbf{D} + \eta_3(D_{11} + D_{33}) \tag{3.17}$$

$$ {}^s\widetilde{S}_{23} = {}^s\widetilde{S}_{32} = \lambda + 2\eta_1 \mathrm{Tr}\mathbf{D} + \eta_3(D_{22} + D_{33}) $$

$$\begin{aligned} {}^s\widetilde{S}_{16} &= {}^s\widetilde{S}_{61} = \eta_3 D_{12} + \tfrac{1}{2}\eta_4 \left[D_{11}b(2a + d) + bdD_{22} \right. \\ &\left. + 2D_{12}\left(b^2 + ad\right)\right] \end{aligned}$$

$$\begin{aligned} {}^s\widetilde{S}_{26} &= {}^s\widetilde{S}_{62} = \eta_3 D_{12} + \tfrac{1}{2}\eta_4 \left[bdD_{11} + D_{22}b(2c + d) \right. \\ &\left. + 2D_{12}\left(b^2 + cd\right)\right] \end{aligned}$$

$$ {}^s\widetilde{S}_{36} = {}^s\widetilde{S}_{63} = \eta_3 D_{12} $$

$$ {}^s\widetilde{S}_{66} = \mu + \eta_2 \mathrm{Tr}\mathbf{D} + \tfrac{1}{2}\eta_4 \left[(D_{11} + D_{22})\left(b^2 + d^2\right) + 4D_{12}bd\right] $$

In the above equations a, b, c and d are simplified functions of the plane rotation angle α and k_1, k_2, which account for the unilateral effect (Kuna-Ciskał and Skrzypek [208]):

$$\begin{aligned} a &= \tfrac{1}{2}(k_1 + k_2) + \tfrac{1}{2}(k_1 - k_2)\cos 2\alpha & b &= \tfrac{1}{2}(k_1 - k_2)\sin 2\alpha \\ c &= \tfrac{1}{2}(k_1 + k_2) - \tfrac{1}{2}(k_1 - k_2)\cos 2\alpha & d &= \tfrac{1}{2}(k_1 + k_2) \end{aligned} \tag{3.18}$$

where the simplified form of the transformation rule (2.55) is used

$$\begin{Bmatrix} \varepsilon_{11}^{e*} \\ \varepsilon_{22}^{e*} \\ \varepsilon_{12}^{e*} \\ \varepsilon_{33}^{e*} \end{Bmatrix} = \begin{bmatrix} a & 0 & b & 0 \\ 0 & c & b & 0 \\ b/2 & b/2 & d & 0 \\ 0 & 0 & 0 & 1 \end{bmatrix} \begin{Bmatrix} \varepsilon_{11}^{e} \\ \varepsilon_{22}^{e} \\ \varepsilon_{33}^{e} \\ \varepsilon_{33}^{e} \end{Bmatrix} \tag{3.19}$$

Thermodynamic-force conjugates $\mathbf{Y}$ and B are defined by (3.6) as follows:

$$\begin{aligned}
Y_{11} &= -\eta_1(\mathrm{Tr}\varepsilon^e)^2 - \eta_2\mathrm{Tr}(\varepsilon^e)^2 - \eta_3(\mathrm{Tr}\varepsilon^e)\varepsilon_{11}^e - \eta_4\Big[(a\varepsilon_{11}^e + b\varepsilon_{12}^e)^2 \\
&\quad + \big(\tfrac{1}{2}b(\varepsilon_{11}^e + \varepsilon_{22}^e) + d\varepsilon_{12}^e\big)^2\Big] \\
Y_{22} &= -\eta_1(\mathrm{Tr}\varepsilon^e)^2 - \eta_2\mathrm{Tr}(\varepsilon^e)^2 - \eta_3(\mathrm{Tr}\varepsilon^e)\varepsilon_{22}^e - \eta_4\Big[(c\varepsilon_{22}^e + b\varepsilon_{12}^e)^2 \\
&\quad + \big(\tfrac{1}{2}b(\varepsilon_{11}^e + \varepsilon_{22}^e) + d\varepsilon_{12}^e\big)^2\Big] \\
Y_{33} &= -\eta_1(\mathrm{Tr}\varepsilon^e)^2 - \eta_2\mathrm{Tr}(\varepsilon^e)^2 - \eta_3(\mathrm{Tr}\varepsilon^e)\varepsilon_{33}^e - \eta_4 k_3^2(\varepsilon_{33}^e)^2 \\
Y_{23} &= 0 \\
Y_{13} &= 0 \\
Y_{12} &= -\eta_3(\mathrm{Tr}\varepsilon^e)^2\varepsilon_{12}^e - \eta_4\big(\tfrac{1}{2}b(\varepsilon_{11}^e + \varepsilon_{22}^e) + d\varepsilon_{12}^e\big)(a\varepsilon_{11}^e + 2b\varepsilon_{12}^e + c\varepsilon_{22}^e)
\end{aligned} \tag{3.20}$$

$$B = K_d\beta \tag{3.21}$$

where

$$\begin{aligned}
\mathrm{Tr}\varepsilon^e &= \frac{(\varepsilon_{11}^e + \varepsilon_{22}^e)\big[2\mu + 2\eta_2\mathrm{Tr}\mathbf{D} + (\eta_3 + 2\eta_4 k_3^2)D_{33}\big]}{\lambda + 2\mu + 2(\eta_1 + \eta_2)\mathrm{Tr}\mathbf{D} + 2(\eta_3 + \eta_4 k_3^2)D_{33}} \\
&\quad - \frac{\eta_3(D_{11}\varepsilon_{11}^e + D_{22}\varepsilon_{22}^e + 2D_{12}\varepsilon_{12}^e)}{\lambda + 2\mu + 2(\eta_1 + \eta_2)\mathrm{Tr}\mathbf{D} + 2(\eta_3 + \eta_4 k_3^2)D_{33}}
\end{aligned} \tag{3.22}$$

Note that the 4×4 secant stiffness matrix $^s\widetilde{\mathbf{S}}(\mathbf{D})$ may further be reduced to 3×3 by eliminating ε_{33}^e from (3.16).

When the general incremental form of the constitutive equations (3.13) is applied to plane stress $d\sigma_{33} = 0$, the following 4×4 tangent stiffness matrix representation $\widetilde{S}_{ij}(\varepsilon^e, \mathbf{D})$ is obtained in terms of the corresponding secant matrix components and the additional terms depending on damage and strain tensors (3.12):

$$\begin{Bmatrix} d\sigma_{11} \\ d\sigma_{22} \\ 0 \\ d\sigma_{12} \end{Bmatrix} = \begin{bmatrix} \widetilde{S}_{11} & \widetilde{S}_{12} & \widetilde{S}_{13} & \widetilde{S}_{16} \\ & \widetilde{S}_{22} & \widetilde{S}_{23} & \widetilde{S}_{26} \\ & & \widetilde{S}_{33} & \widetilde{S}_{36} \\ & & & \widetilde{S}_{66} \end{bmatrix} \begin{Bmatrix} d\varepsilon_{11}^e \\ d\varepsilon_{22}^e \\ d\varepsilon_{33}^e \\ d\gamma_{12}^e \end{Bmatrix} \tag{3.23}$$

Failure criterion

When the matrix representation is used, the constitutive law of the elasto-damage material can be written in the incremental form as follows:

$$
\begin{Bmatrix} d\sigma_{11} \\ d\sigma_{22} \\ d\sigma_{33} \\ d\sigma_{23} \\ d\sigma_{31} \\ d\sigma_{12} \end{Bmatrix} =
\begin{bmatrix}
\frac{\partial^2 \Psi}{\partial \varepsilon_{11}^2} & \frac{\partial^2 \Psi}{\partial \varepsilon_{11} \partial \varepsilon_{22}} & \frac{\partial^2 \Psi}{\partial \varepsilon_{11} \partial \varepsilon_{33}} & \frac{\partial^2 \Psi}{\partial \varepsilon_{11} \partial \gamma_{23}} & \frac{\partial^2 \Psi}{\partial \varepsilon_{11} \partial \gamma_{31}} & \frac{\partial^2 \Psi}{\partial \varepsilon_{11} \partial \gamma_{12}} \\
& \frac{\partial^2 \Psi}{\partial \varepsilon_{22}^2} & \frac{\partial^2 \Psi}{\partial \varepsilon_{22} \partial \varepsilon_{33}} & \frac{\partial^2 \Psi}{\partial \varepsilon_{22} \partial \gamma_{23}} & \frac{\partial^2 \Psi}{\partial \varepsilon_{22} \partial \gamma_{31}} & \frac{\partial^2 \Psi}{\partial \varepsilon_{22} \partial \gamma_{12}} \\
& & \frac{\partial^2 \Psi}{\partial \varepsilon_{33}^2} & \frac{\partial^2 \Psi}{\partial \varepsilon_{33} \partial \gamma_{23}} & \frac{\partial^2 \Psi}{\partial \varepsilon_{33} \partial \gamma_{31}} & \frac{\partial^2 \Psi}{\partial \varepsilon_{33} \partial \gamma_{12}} \\
& & & \frac{\partial^2 \Psi}{\partial \gamma_{23}^2} & \frac{\partial^2 \Psi}{\partial \gamma_{23} \partial \gamma_{31}} & \frac{\partial^2 \Psi}{\partial \gamma_{23} \partial \gamma_{12}} \\
& & & & \frac{\partial^2 \Psi}{\partial \gamma_{31}^2} & \frac{\partial^2 \Psi}{\partial \gamma_{31} \partial \gamma_{12}} \\
& & & & & \frac{\partial^2 \Psi}{\partial \gamma_{12}^2}
\end{bmatrix}
\begin{Bmatrix} d\varepsilon_{11}^e \\ d\varepsilon_{22}^e \\ d\varepsilon_{33}^e \\ d\gamma_{23}^e \\ d\gamma_{31}^e \\ d\gamma_{12}^e \end{Bmatrix}
$$

$$(3.24)$$

In order to introduce the general *failure criterion* the Drucker's material stability postulate is adopted:

$$d\sigma_{ij} d\varepsilon_{ij}^e > 0 \tag{3.25}$$

hence

$$\frac{\partial^2 \Psi}{\partial \varepsilon_{ij} \partial \varepsilon_{kl}} d\varepsilon_{ij} d\varepsilon_{kl} = H_{ijkl} d\varepsilon_{ij} d\varepsilon_{kl} > 0 \tag{3.26}$$

The quadratic form $(\partial^2 \Psi / \partial \varepsilon_{ij} \partial \varepsilon_{kl}) d\varepsilon_{ij} d\varepsilon_{kl}$ must be positive definite for arbitrary values of the components $d\varepsilon_{ij}^e$, hence, eventually the condition (3.26) requires that the *local Hessian matrix* $[\mathbf{H}]$ be positive definite (Chen and Han [42]). According to the Sylvester criterion the symmetric matrix $[\mathbf{H}]$ of the nth order is positive definite if and only if

$$\det [\mathbf{H}_k] > 0 \qquad k = 1, 2, ..., n \tag{3.27}$$

where $[\mathbf{H}_k]$ is the $(k \times k)$ minor of the matrix $[\mathbf{H}]$ (Björck and Dahlquist [23]).

Implementation of (3.27) may also be performed by use of the Cholesky decomposition, which is numerically stable. For a symmetric and positive definite matrix $[\mathbf{H}]$ the Cholesky decomposition constructs a lower triangular matrix $[\mathbf{A}]$, the components of which are obtained as follows:

$$A_{ii} = \left(H_{ii} - \sum_{k=1}^{i-1} A_{ik}^2 \right)^{\frac{1}{2}} \tag{3.28}$$

$$A_{ji} = \frac{1}{A_{ii}} \left(H_{ij} - \sum_{k=1}^{i-1} A_{ik} A_{jk} \right) \qquad j = i+1, i+2, ...6 \tag{3.29}$$

Finally, the failure criterion may be furnished form (3.27) and (3.28) as

$$\det [\mathbf{A}_i] = 0 \qquad \text{or} \qquad |H_{ii}| = \sum_{k=1}^{i-1} A_{ik}^2 \tag{3.30}$$

3.1.2 Numerical Simulation of Damage and Fracture in Plane Concrete Structure in the Presence of an Initial Crack using the Local Approach

The local tangent stiffness matrix $\widetilde{S}_{ij}(\varepsilon^{\mathrm{e}}, \mathbf{D})$ is used for the quasi-Newton algorithm for first iteration step of solving the nonlinear equation (3.13) as long as the local failure criterion (3.30) is met. The stiffness of the element in the FE mesh that has failed is next reduced to zero. As a consequence, the failed element is completely released from stress and the appropriate stress redistribution occurs in the neighbouring elements to ensure global equilibrium. Note that the above failure criterion (3.27) assumes the brittle failure mechanism. However, when a broader class of materials is considered, a post-peak softening regime can also be admitted, that would result in strain localization and a smooth stiffness drop in elements that fail.

By the use of *Local Approach to Fracture* LAF, based on FEM and CDM, the crack is modelled as an assembly of failed elements in the mesh. Subsequent elements in the FE mesh that have failed (along the crack) are released from stresses and the appropriate redefinition of the global stiffness of the structure is made. The procedure is continued as long as the overall *fracture mechanism* of the structure is reached. All the numerical calculations are performed by use of the ABAQUS system.

The LAF method is capable of predicting fracture initiation and the ultimate fracture pattern, as well as evaluating the limit load that corresponds to the considered fracture mechanism in a structure. The crack growth stage and the ultimate crack pattern are, in general, mesh-dependent. To mitigate the effect of element size and shape, additional regularization procedures are required.

In the examples presented below, regularization has not been used. The results obtained for crack prediction exhibit the mesh effect, so that a nonlocal treatment has to be involved in order to ensure convergence. This procedure is examined and presented in the following Sect. 3.1.3.

Monotonic tension – kinked crack

Following the paper by Kuna-Ciskał and Skrzypek [208], the plane-stress concrete structure with a pre-load crack of infinitely small width, inclined at an angle of 45° to the tension direction is considered. The structure is subject to uniform tension at the top and bottom edges, and is free to move at all sides, Fig. 3.1. The material constants correspond to a high strength concrete (Murakami and Kamiya [162]):

$$
\begin{aligned}
&E = 21,4\,[\mathrm{GPa}] \quad \nu = 0.2 \qquad\qquad \eta_1 = -400\,[\mathrm{MPa}] \\
&\eta_2 = -900\,[\mathrm{MPa}] \;\; \eta_3 = 100\,[\mathrm{MPa}] \quad \eta_4 = -23500\,[\mathrm{MPa}] \\
&\zeta = 0.1 \qquad\qquad K_{\mathrm{d}} = 0.04\,[\mathrm{MPa}] \;\; B_0 = 0,0026\,[\mathrm{MPa}]
\end{aligned}
\tag{3.31}
$$

The monotonically increasing tensile load q causes the pre-load crack of "zero" width in the pre-load state, to open. Stress concentration at the zones neighbouring the crack tips is accompanied by cumulation of the D_{11} damage component in the element where the secondary kinked crack starts to open when the failure criterion (3.30) is met locally. The magnitudes and directions of the principal stress components in elements neighbouring the crack tip are sketched in Fig. 3.1 just before the instant of the secondary kinked crack opening. The magnified square zone neighbouring the pre-load crack tip is shown

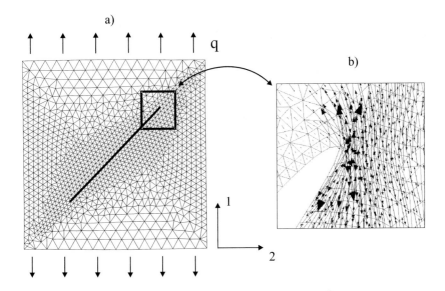

Fig. 3.1. A plane stress concrete structure with pre-load inclined crack, subjected to tension in the direction 1: (a) the geometry and mesh, (b) maximal principal stress distribution surrounding the pre-load crack tip; the arrows represent the actual principal stresses

in Fig. 3.1b, where a magnification factor for displacements equal to 100 is used for clarity. With increasing tensile load, consecutive elements fail, to ultimately form a *kinked crack* that splits the concrete specimen approximately perpendicularly to the tension direction, as shown in Fig. 3.2a, b, c. Note that the elements neighbouring the crack face are released from stresses following the kinked crack growth. In the final stage the anisotropic damage growth and fracture mechanism in a specimen causes the overall stiffness of the structure to drop drastically. The predicted critical load for crack initiation q_1^f differs from that of ultimate failure q_u^f by an amount approximately 15%.

A qualitatively similar kinked crack in a brittle rock-like specimen was predicted by Basista [15], where the micromechanical damage model was used on the microlevel.

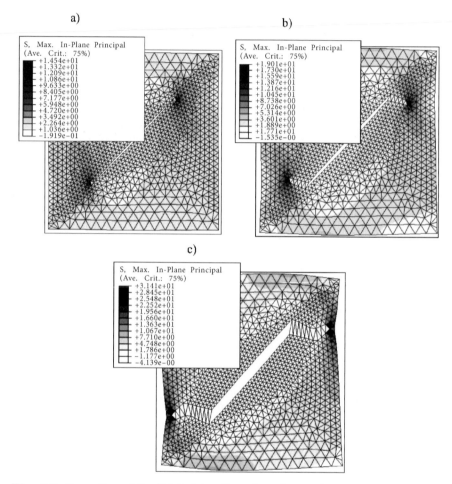

Fig. 3.2. Formation of the "kinked-type" crack under tension; maximum principal stress distribution at different stages of crack propagation

Monotonic compression – wing crack

To illustrate the failure mechanism under compression, consider the plane concrete structure with pre-load crack inclined at an angle of about 70° to the compression direction. Uniform compression is applied at top and bottom edges, and the structure is free to move at all sides. No confinement is applied at specimen sidewalls (Fig. 3.3) and a frictionless pre-load crack is assumed for simplicity.

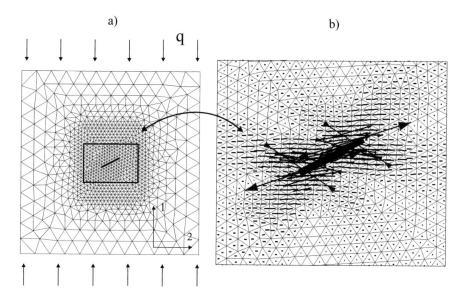

Fig. 3.3. A plane stress concrete structure with pre-load inclined crack, subjected to compression in the direction 1: (a) the geometry and mesh, (b) maximum principal stress distribution surrounding the pre-load crack tip

Contact at the pre-load crack edges was defined by identifying and pairing potential contact surfaces. To define a sliding interface between two surfaces, one of the surfaces (the "slave" surface) is covered with 3-node contact elements. The second surface (the "master" surface) is defined by a series of nodes ordered as the segments of the slide line. For each node on the slave surface the closest point on the master surface is looked for, where the master surface's normal passes through the node on the slave surface. The interaction is then discretized between the point on the master surface and the slave node (cf. ABAQUS *Theory Manual* [1]).

FEM analysis, performed at the end of the first phase when sliding occurs on pre-load crack faces with no crack length change, shows the maximum tensile stress vectors at both crack tips. The maximum tensile stress causes the secondary wing crack to open (Fig. 3.4a), when the failure criterion (3.30) is reached. Next the stiffness (and stress) in the failed element is reduced to zero. During the next loading steps subsequent elements in the FE mesh fail, to eventually form two straight *wing cracks* growing in the compressive load direction (Fig. 3.4b).

After a number of loading steps a slight change in the crack path was observed, indicating a mixed type tensile–shear mechanism of crack growth

a)   b)

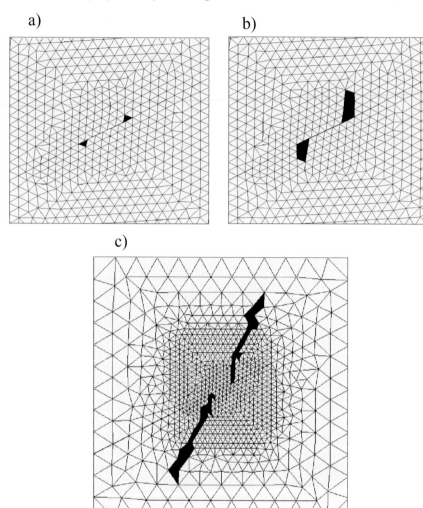

c)

Fig. 3.4. Formation of the "wing-type" crack under compression; subsequent stages of the macrocrack growth

(Fig. 3.4c). However, the mesh effect may also be more significant there. When the secondary crack splits the structure, the zones neighbouring the crack faces are gradually released from stresses, whereas the front of the maximum tensile stress propagates outwards and the wing crack length increases (Fig. 3.5a, b, c). The predicted wing cracks in a concrete specimen under compression do not exhibit curvilinear shape; they start from the pre-load crack tips in a straight line manner, growing in a direction approximately parallel to the loading axis (Fanella and Krajcinovic [72]).

a)   b)

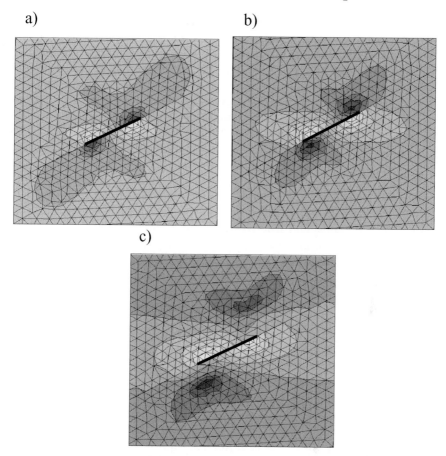

c)

Fig. 3.5. Maximum principal stress distribution at several stages of "wing-type" crack growth under compression

3.1.3 Nonlocal Formulation of the Constitutive Equations

In this section, following the paper by Egner *et al.* [64], the new *nonlocal integral-type model* is developed and examined for simulating anisotropic, localized damage and the prediction of failure modes in the double-notched plane-stress concrete specimen used in the Nooru-Mohammed experiment (Nooru-Mohammed [167]).

The use of the classical local constitutive models is insufficient for problems where a strong strain softening effect occurs. In the case of inelastic material behaviour two dissipative processes are responsible for the strain softening: (visco)plasticity and/or damage (Hansen and Schreyer [94], Abu Al-Rub and Voyiadjis [6]). In the case of *Local Models* (LM) the stress at a given point is assumed to be uniquely determined by the strain history at this point only. However, as the (visco)plasticity and damage frequently localize over narrow zones of the continuum, statistical homogeneity in the representative volume element is lost. Hence, the *characteristic length scale* has to be introduced in the *Nonlocal Model* (NL) in order to account for the influence of an internal state variable also at neighbouring points. From a computational point of view, in the case of localized phenomena, ill-posedness of the boundary value problem and mesh sensitivity of finite element computations are met. Incorporation of viscosity retains the ellipticity of the problem, such that the well-possedness is preserved, because viscosity implicitly introduces length-scale measures that reduce the strain and damage localization (Wang et al. [231], Dornowski and Perzyna [58], Glema et al. [89]). A theoretical framework for nonlocal combined viscosity and gradient-enhanced theory of anisotropic thermo-plastic-damage mechanism was recently developed in Abu Al-Rub and Voyiadjis [7] and implemented in the explicit finite element code ABAQUS in Voyiadjis and Abu Al-Rub [230].

However, if (visco)plasticity is not accounted for, some computational *localization limiters* should be used, provided by the concept of nonlocal *weighted averaging* (Bažant [17], Pijaudier-Cabot and Bažant [180], Bažant and Pijaudier-Cabot [19], Jirasek [120], Comi [49], Comi and Perego [50], Voyiadjis and Abu Al-Rub [229]).

If $f(\mathbf{x})$ is a local field in a volume, the corresponding nonlocal field is defined as

$$\bar{f}(\mathbf{x}) = \frac{1}{V} \int\limits_V h(\mathbf{x}, \boldsymbol{\xi}) f(\boldsymbol{\xi}) \mathrm{d}\boldsymbol{\xi} \qquad (3.32)$$

where $h(\mathbf{x}, \boldsymbol{\xi})$ is a monotonically decreasing weight function, defined in such a way that a uniform field is not altered by it, and

$$V(\boldsymbol{\xi}) = \int\limits_V h(r) \mathrm{d}V, \qquad r = |\mathbf{x} - \boldsymbol{\xi}| \qquad (3.33)$$

As the weight function $h(r)$ the *Gauss distribution function*

$$h(r) = \exp\left(-\frac{r^2}{2l^2}\right) \qquad (3.34)$$

or the *bell-shape function*

$$h(r) = \begin{cases} \left(1 - \dfrac{r^2}{R^2}\right)^2 & 0 \leq r \leq R \\ 0 & R \leq r \end{cases} \qquad (3.35)$$

are frequently used. In the above l is the *internal length* of the *nonlocal continuum*, whereas R is called the *interaction length*, which is related, but not equal, to the internal length l. If the orientation of the material fibres is considered, a more complicated averaging operator might be used, where not, only the distance between points $\mathbf{x}$ and $\boldsymbol{\xi}$, but also the orientation of principal axes at these points, are accounted for (Bažant [18]).

The essential point is how to properly choose the internal length parameter of the nonlocal continuum. This length may be assessed by the experimental comparison of energy in a specimen where damage is constrained to remain diffuse, and another one where damage localizes to yield a single crack (Mazars and Pijaudier-Cabot [156, 157]). It may also be established from the maximum size of the aggregate in concrete d_a, such that $l \approx 3d_\mathrm{a}$ holds (also Saouridis and Mazars [196]). Some particular suggestions can be found in the comparative study on different models for concrete: local (Ottosen) or nonlocal (the nonlocal damage and the gradient plasticity) (di Prisco et al. [56]). The internal length values are thus set to $l \approx 5$ mm for the gradient plasticity model, and to $l \approx 8$ mm for the nonlocal damage model. On the other hand, Comi and Perego [50] use for the nonlocal concrete model, the value $l \approx 1.1$ mm.

Nonlocal formulation of the damage dissipation potential and evolution equations

The local state equations combined with the local evolution law for $\dot{\mathbf{D}}$ and $\dot{\beta}$ (see Sect. 3.1.1) are not capable of predicting the damage evolution in the case of the localized damage and strain fields because of a spurious mesh effect. In general, to avoid the singularity of $\mathbf{Y}$ at the crack tip when the mesh size tends to zero, the *Cut–off Algorithm* (CA) may optionally be used in the neighbourhood of a crack tip, according to the scheme (Skrzypek et al. [209, 210]):

$$\widehat{\mathbf{Y}} = k\mathbf{Y}, \quad k = \begin{cases} 1 & \text{if } Y_\mathrm{eq} \leq Y_\mathrm{u} \\ Y_\mathrm{u}/Y_\mathrm{eq} & \text{if } Y_\mathrm{eq} > Y_\mathrm{u} \end{cases} \tag{3.36}$$

where the cut-off factor k is defined as follows:

$$k = Y_\mathrm{u}/Y_\mathrm{eq} = (B_0 + B)/Y_\mathrm{eq} \tag{3.37}$$

and B_0 denotes the initial damage threshold. The new variable $\widehat{\mathbf{Y}}$ is next subjected to the nonlocal treatment (NA) $\overline{\widehat{\mathbf{Y}}}$, according to the following formula (Pijaudier-Cabot and Bažant [180], Comi and Perego [50]):

$$\overline{\widehat{\mathbf{Y}}}(\mathbf{x}) = \frac{\int\limits_{\Omega_\mathrm{d}} \widehat{\mathbf{Y}}(\boldsymbol{\xi})\varphi(\mathbf{x},\boldsymbol{\xi})\mathrm{d}\Omega_\mathrm{d}}{\int\limits_{\Omega_\mathrm{d}} \varphi(\mathbf{x},\boldsymbol{\xi})\mathrm{d}\Omega_\mathrm{d}} \tag{3.38}$$

$$\varphi(\mathbf{x},\boldsymbol{\xi}) = \exp\left[-\left(d(\mathbf{x},\boldsymbol{\xi})/l\right)^2\right] \tag{3.39}$$

where $\varphi(\mathbf{x},\boldsymbol{\xi})$ is a *weight function* and l is the *internal length of the nonlocal continuum*.

The damage dissipation potential in the space of *nonlocal force conjugates* $\left\{\widehat{\overline{\mathbf{Y}}}, -\overline{B}\right\}$ is assumed to be in analogous form to that in the local space $\{\mathbf{Y}, -B\}$ (3.7):

$$\overline{F}^{\mathrm{d}}(\widehat{\overline{\mathbf{Y}}}, \overline{B}) = \widehat{\overline{Y}}_{\mathrm{eq}} - (B_0 + \overline{B}) = 0, \quad \widehat{\overline{Y}}_{\mathrm{eq}} = \sqrt{0,5\widehat{\overline{\mathbf{Y}}} : \mathbf{L} : \widehat{\overline{\mathbf{Y}}}} \quad (3.40)$$

In Eq. (3.40) $\overline{B}$ denotes the nonlocal damage force conjugate of the nonlocal damage hardening variable $\overline{\beta}$. The *nonlocal evolution equations* for $\dot{\overline{\beta}}$ and $\dot{\overline{\mathbf{D}}}$ are finally established from the normality rule:

$$\dot{\overline{\mathbf{D}}} = \dot{\overline{\lambda}}^{\mathrm{d}} \frac{\partial \overline{F}^{\mathrm{d}}}{\partial \widehat{\overline{\mathbf{Y}}}}, \quad \dot{\overline{\beta}} = \dot{\overline{\lambda}}^{\mathrm{d}} \quad (3.41)$$

where the consistency condition is used to calculate nonlocally defined scalar damage multiplier $\dot{\overline{\lambda}}_{\mathrm{d}}$

$$\dot{\overline{F}}^{\mathrm{d}} = 0 = (\partial \overline{F}^{\mathrm{d}}/\partial \widehat{\overline{\mathbf{Y}}}) : \dot{\widehat{\overline{\mathbf{Y}}}} + (\partial \overline{F}^{\mathrm{d}}/\partial \overline{B})\dot{\overline{B}} \quad (3.42)$$

$$\dot{\overline{\lambda}}^{\mathrm{d}} = \alpha \frac{\mathbf{L} : \widehat{\overline{\mathbf{Y}}}}{2K_{\mathrm{d}}\widehat{\overline{Y}}_{\mathrm{eq}}} : \dot{\widehat{\overline{\mathbf{Y}}}} \quad (3.43)$$

As in (3.9), a factor $\alpha = 1$ or $\alpha = 0$ is used for active or passive damage growth, respectively.

Incremental formulation of the nonlocal model and the failure criterion

In the case of a nonlocal continuum the new, nonlocal variables $\overline{\mathbf{Y}}$ and $\overline{B}$ or, alternatively $\widehat{\overline{\mathbf{Y}}}$ and $\overline{B}$, are used in the evolution equations, instead of the local ones $\mathbf{Y}$ and B. Hence, the internal damage variables are also defined in a nonlocal sense, e.g. $\overline{\mathbf{D}}$ and $\overline{\beta}$. Other state variables, $\boldsymbol{\sigma}$ and $\boldsymbol{\varepsilon}^{\mathrm{e}}$ are not subjected to the nonlocal averaging, however, the locally defined stiffness matrix ${}^{\mathrm{s}}\mathbf{S}(\mathbf{D})$ has to be replaced by the new, nonlocally prescribed secant matrix ${}^{\mathrm{s}}\overline{\overline{\mathbf{S}}}(\overline{\mathbf{D}})$, also called the *nonlocal secant stiffness*, that accounts for damage nonlocality. Then, the new *nonlocal effective tangent stiffness* matrix $\overline{\overline{\mathbf{S}}}(\boldsymbol{\varepsilon}^{\mathrm{e}}, \overline{\mathbf{D}})$ has to be established to yield the incremental constitutive equation in a nonlocal fashion

$$\mathrm{d}\boldsymbol{\sigma} = {}^{\mathrm{s}}\overline{\overline{\mathbf{S}}} : \mathrm{d}\boldsymbol{\varepsilon}^{\mathrm{e}} + \boldsymbol{\varepsilon}^{\mathrm{e}} : \frac{\partial {}^{\mathrm{s}}\overline{\overline{\mathbf{S}}}}{\partial \overline{\mathbf{D}}}\mathrm{d}\overline{\mathbf{D}} \quad (3.44)$$

Finally, applying (3.41) to obtain nonlocal damage increments $\mathrm{d}\overline{\mathbf{D}}$ on $\mathrm{d}\boldsymbol{\varepsilon}^{\mathrm{e}}$, the following *nonlocal incremental state equation* is derived (Egner *et al.* [64]):

$$d\boldsymbol{\sigma} = \left[{}^{s}\overline{\overline{\mathbf{S}}} + \alpha\varepsilon^{e} : \frac{\partial\, {}^{s}\overline{\overline{\mathbf{S}}}}{\partial\overline{\mathbf{D}}} : \frac{\partial\overline{\mathbf{D}}}{\partial\varepsilon^{e}} \right] : d\varepsilon^{e} = \overline{\overline{\mathbf{S}}}\left(\varepsilon^{e}, \overline{\mathbf{D}}\right) : d\varepsilon^{e} \qquad (3.45)$$

or, when the matrix–vector notation, is used

$$\{d\boldsymbol{\sigma}\} = \left[\overline{\overline{\mathbf{S}}}\left(\varepsilon^{e}, \overline{\mathbf{D}}\right)\right]\{d\varepsilon^{e}\} \qquad (3.46)$$

The square bracket in (3.46) represents the new nonlocal effective tangent stiffness $\overline{\overline{\mathbf{S}}}\left(\varepsilon^{e}, \overline{\mathbf{D}}\right)$ that follows the damage nonlocality $\overline{\mathbf{D}}$.

In order to introduce the general failure criterion in a nonlocal sense, Drucker's material stability postulate is adopted:

$$\left(\partial^{2}\overline{\Psi}/\partial\varepsilon_{ij}\partial\varepsilon_{kl}\right)d\varepsilon_{ij}d\varepsilon_{kl} = \overline{H}_{ijkl}d\varepsilon_{ij}d\varepsilon_{kl} > 0 \qquad (3.47)$$

when the nonlocally defined quadratic form $\left(\partial^{2}\overline{\Psi}/\partial\varepsilon_{ij}\partial\varepsilon_{kl}\right)d\varepsilon_{ij}d\varepsilon_{kl}$, must be positive definite for arbitrary values of the components $d\varepsilon_{ij}$. Hence, eventually the *nonlocal Hessian matrix* $\left[\overline{\mathbf{H}}\right]$ must be positive definite, such that the following holds

$$\det\left[\overline{\mathbf{H}}_{k}\right] > 0 \qquad k = 1, 2, ... n \qquad (3.48)$$

3.1.4 Numerical Simulation of the Nonlocal Damage and Fracture in a Double-Notched Plane-Stress Concrete Specimen

In what follows, a proper characteristic length is numerically assessed from the simulation tests on damage and fracture prediction in the double-notched specimen (see Fig. 3.6). Different values of characteristic lengths for concrete are numerically tested, ranging from 0 mm (local) to 20 mm, to finally assess the value $l = 7.5$ mm as an "optimal" one, to preserve the characteristic damage incubation and ultimate localized failure prediction without violating stability and mesh convergence.

The double-edge notched plane-stress specimen follows that used in the Nooru-Mohammed experiment (Nooru-Mohammed [167], Nooru-Mohammed *et al.* [168]). It allows for the analysis of various combinations of shear and tension under displacement control. The model was investigated in di Prisco et al. [56] to simulate fracture by means of the three approaches: the *local model*, the *gradient plasticity model* and the *nonlocal damage model*. In what follows, in order to determine the Mode I crack growth under tension, the shear component was excluded ($\delta_{t} = 0$). The material data for a high strength concrete that describe the basic local Murakami–Kamiya model are taken after Murakami and Kamiya [162] (3.31).

Assuming uniform normal displacement δ_{n} applied at the top of the specimen shown in Fig. 3.6 ($\delta_{t} = 0$), a complete process of damage growth and fracture is simulated until the ultimate failure of the specimen is predicted. A combined non-symmetric tension/shear failure mode is developed due to the non-symmetric boundary conditions used (Fig. 3.6). Two zones of failed

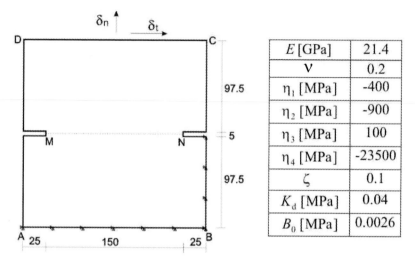

Fig. 3.6. Double-edge notched specimen configuration (di Prisco *et al.* [56]) and material data (Murakami and Kamiya [162])

elements where the ellipticity is lost (checking the nonlocal Hessian matrix $\overline{\mathbf{H}}$ (3.47)) spread inwards in opposite directions from the notches as long as the ultimate fracture mechanism is not achieved. Releasing the consecutive failed elements from stresses results in stress redistribution in neighbouring (non-failed) elements. The distribution of stresses along MN line (Fig. 3.6) becomes more and more non-uniform, finally yielding a strong stress localization in front of two failed zones that come together when the ultimate fracture is met.

To check the influence of the characteristic length of the nonlocal continuum (3.39) on the damage evolution and fracture processes the following values of $l = 2.5, 5.0, 7.5, 10.0, 20.0$ mm are tested. The range of values is taken on the basis of di Prisco *et al.* [56], where for the nonlocal damage model $l \approx 8$ mm is adopted. The finite element size must be less than the characteristic length to make the nonlocal approach active, so the rectangular mesh 2.5 mm×2.5 mm is adopted here (Fig. 3.7b). Other meshes, shown in Fig 3.7a and Fig. 3.6c are defined for the convergence test.

The effect of increasing value of l on $\overline{Y}_{eq}$ is shown in Figs 3.8 and 3.9. The characteristic length defines the area over which Y_{eq} is averaged, according to (3.38–3.39). The bigger l – the larger the area, the more balanced and lower values of averaged variable around the present integration point (Fig. 3.8), and the less advanced damage and fracture process at a chosen level of load (Fig. 3.9). The increasing value of l clearly makes the fracture progress slower, as shown in Fig. 3.9a–c. Depending on the value of l, at the same postcritical loading (step 54), different advances of fracture are met. In other words, increasing l results in increasing both critical displacements as

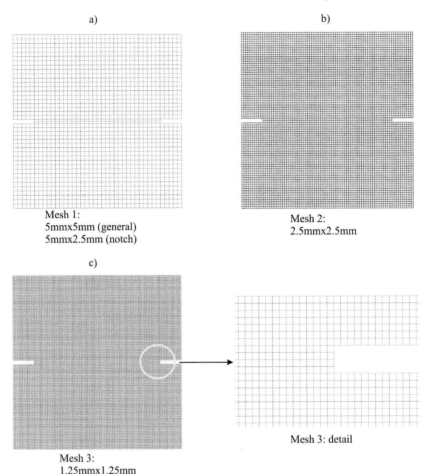

a)

Mesh 1:
5mmx5mm (general)
5mmx2.5mm (notch)

b)

Mesh 2:
2.5mmx2.5mm

c)

Mesh 3:
1.25mmx1.25mm

Mesh 3: detail

Fig. 3.7. FEM meshes for convergence tests

shown in Fig. 3.10. The value of the "incubation" displacement (Fig. 3.10) denotes the displacement of the edge DC (Fig. 3.6) at the instant of macro-crack appearance, while "fracture" displacement means the displacement of the same edge at the instant of overall failure of the structural element. It is observed that for $l \leq 7.5$ mm the second critical displacement at fracture becomes (almost) nonsensitive to the characteristic length size.

Simulation of crack growth by the nonlocal approach presented here depends on the characteristic length of a nonlocal continuum (Nooru-Mohammed et al. [168], Skrzypek et al. [209, 210]). When the *nonlocal approach* (NA) is used, an increasing width of crack is observed when increasing the *characteristic length*. At the same time, the direction of macrocracking changes: the lower l the more crack shape tends to the straight shape (see Fig. 3.11).

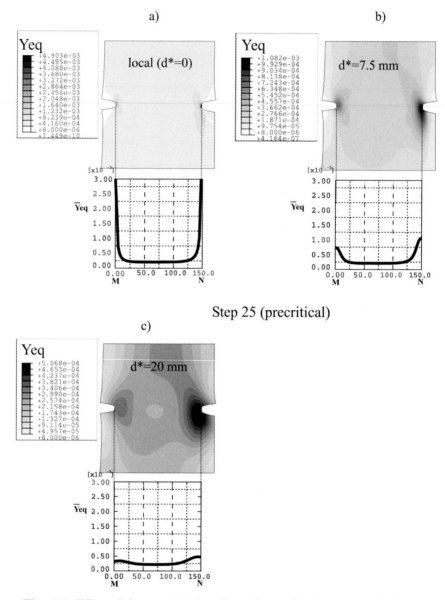

Fig. 3.8. Effect of characteristic length on Y_{eq} localization – precritical stage

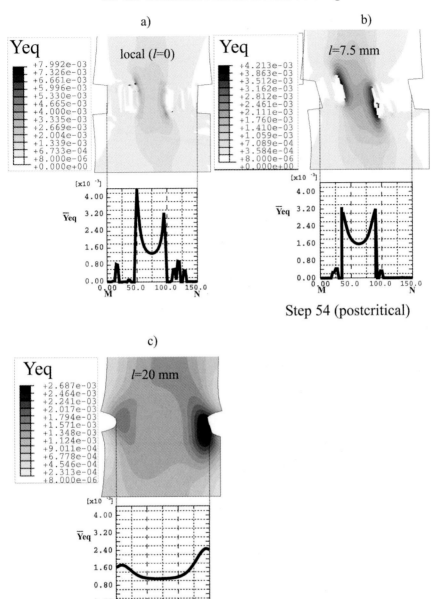

Fig. 3.9. Effect of characteristic length on Y_{eq} localization – postcritical stage

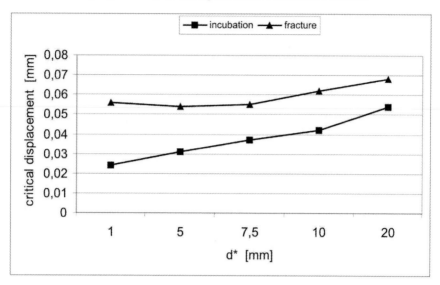

Fig. 3.10. Effect of the characteristic length on critical displacements: incubation and failure

Let us consider the mesh-dependence of the numerical simulation of damage and fracture. The local approach is element shape dependent. For different finite element shapes different crack patterns are usually obtained. Namely, when the local approach is used, the crack may propagate from the current failed element to the neighbouring one, the edge of which is shared with the corresponding edge of the current failed element. In other words, the element shape restricts the possible directions of crack propagation (Murakami and Liu [161], Kuna-Ciskał [136]). By contrast, the *nonlocal approach* presented in this paper allows one to obtain the crack pattern independent of the element shape (Fig. 3.12).

Finally, a *convergence test* was performed for different element sizes. The results shown in Fig. 3.13 proved the capability of the proposed nonlocal model to properly describe both precritical damage evolution during the incubation period, as well as the postcritical crack growth period. It is shown that the use of an additional localization limiter, the cut-off algorithm (CA), is not necessary to meet convergence. The use of the single localization limiter by nonlocal averaging (NA) only is capable of the convergent prediction of both critical displacements, at crack initiation and at fracture.

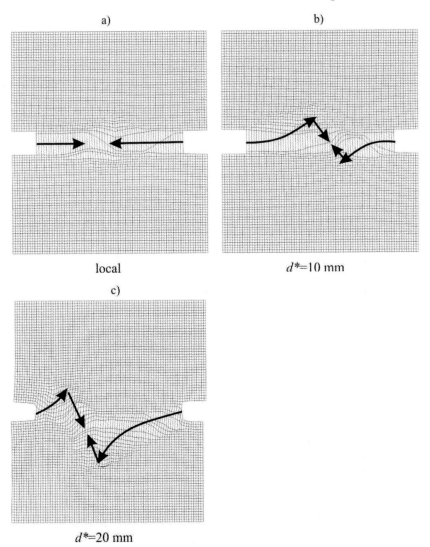

Fig. 3.11. Effect of characteristic length l on mesh deformation (crack pattern) at failure

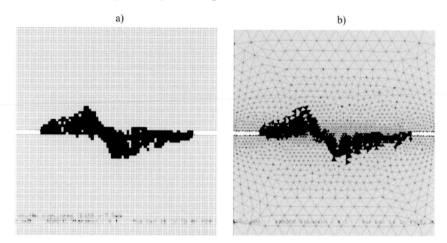

Fig. 3.12. Mesh shape dependence test ($l = 7.5$ mm): (a) regular rectangular mesh, (b) irregular triangular mesh

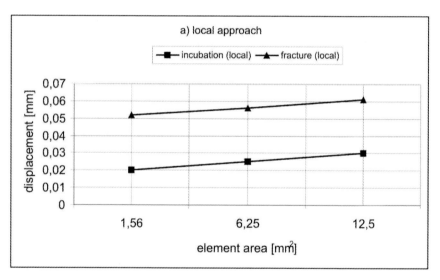

Fig. 3.13. Convergence tests with no localization limiters (local approach)

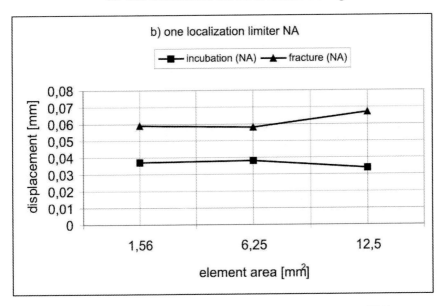

Fig. 3.14. Convergence tests with one localization limiter (NA)

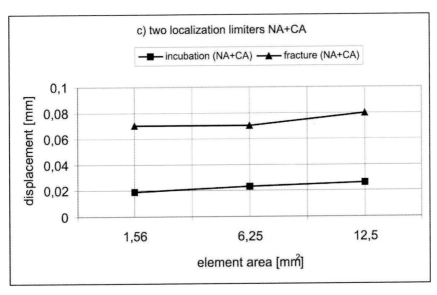

Fig. 3.15. Convergence tests with two localization limiters (NA+CA)

3.1.5 Numerical Simulation of Fracture in a Double-Notched Plane-Stress Concrete Specimen by use of the Nonlocal Equivalent Crack Concept

Introduction

From the mechanical point of view a macrocrack can be described as a discontinuity in the displacement field. The available approaches to model such discontinuities may be classified into two groups:

- The *Fracture Mechanics* (FM) approach. This can be applied once a crack has been initiated, but when the localization and direction of propagation of a macrocrack is not known the method of fracture mechanics cannot be used.
- The *Continuum Damage Mechanics* (CDM) approach. This offers the possibility to predict the location of the expected macrocrack. Taking into account strain softening in the constitutive equations leads to localization effects, which allow one to model the crack as a zone of high gradients of stiffness and strength in which the critical conditions of damage have been reached (Lemaitre [146]). However, the localization phenomena lead also to the ill-posedness of the boundary value problem and mesh dependence (see Chapt. 3.1.4).

There have been attempts to build a bridge between fracture and damage mechanics. One of them is called the *Strong Discontinuity Approach* (SDA), which is based on the standard local continuum constitutive equations (including the effect of strain softening), but additionally, strong discontinuity in kinematics is considered (the strains are unbounded Dirac delta functions) (Oliver *et al.* [172], Wells and Sluys [235]).

Another approach is the *equivalent crack concept*, which allows one to pass from a damage field to a fracture problem and, conversely, a damage zone can be determined which is equivalent to a crack (Mazars and Pijaudier-Cabot [157]). The equivalent crack approach is based on a thermodynamics formalism, which deals with energy considerations and, therefore, it is possible to relate nonlocally defined damage variables (Sect. 3.1.3) and *global fracture variables*. The equivalence of energy dissipation (which must be thermodynamically acceptable) gives the following relation between *nonlocally defined energy release rate* $\overline{\mathbf{Y}}$, and global *fracture energy release rate* G

$$\int_V (-\overline{\mathbf{Y}}\dot{\mathbf{D}} + \overline{B}\dot{\beta})\mathrm{d}V = -G\dot{A}_{\mathrm{eq}} \tag{3.49}$$

where V is the overall volume of the structure. The critical condition of crack propagation, from local fracture mechanics, is $-G = G_c$. Then Eq. (3.49) gives the rate of equivalent crack area $\dot{A}_{\mathrm{eq}}$ to a given evolution of damage $\left\{\dot{\mathbf{D}}, \dot{\beta}\right\}$

$$\dot{A}_{eq} = \frac{\int\limits_{V}(-\overline{\mathbf{Y}}\dot{\mathbf{D}} + \overline{B\beta})\mathrm{d}V}{G_c} \tag{3.50}$$

Considering the complete evolution of damage variables from zero to the current level the equivalent crack area is obtained as

$$A_{eq} = \frac{\int\limits_{V}\int\limits_{0}^{\beta(\mathbf{x})}(-\overline{\mathbf{Y}}\dot{\mathbf{D}} + \overline{B\beta})\mathrm{d}\beta\mathrm{d}V}{G_c} \tag{3.51}$$

where A_{eq} is the current area of equivalent crack, the propagation of which consumed the same energy as the energy consumed during the evolution of damage in the structural element (Mazars and Pijaudier-Cabot [157]).

Numerical procedure

Numerical analysis of the equivalent crack propagation is performed using ABAQUS/Standard. The Murakami–Kamiya constitutive model presented in Sect. 3.1.1 and the failure criterion (3.51) are implemented into the ABAQUS finite element code via a user-supplied procedure that defines non-standard material properties. Iteration of the global equilibrium of a system is performed by ABAQUS using the Newton–Raphson method. The applied scheme is iteration-independent, and all variables are only updated at the end of an increment step after convergence is achieved. The task to be performed by the user-supplied routine is to integrate physical relations at a point level (Gauss point of an element) when starting from a known equilibrium state and for a total strain increment given in each iteration. The output information comprizes stress and all other state variables updated (integrated) by the end of the iteration increment as well as the local stiffness matrix. Here, the integration is performed with a forward Euler scheme. The derivatives (stiffness) are known at start point and kept constant during the increment. Material degradation and failure lead to severe convergence difficulties in implicit analysis programs such as ABAQUS/Standard, therefore the nonlocal integral approach (see Sect. 3.1.3) is applied.

Numerical example: geometry, loading and FE mesh

The plane-stress structure corresponding to a compact tension specimen described in Mazars and Pijaudier-Cabot [157] is shown in Fig. 3.16. Half of the structure (due to symmetry) is divided into 4908 triangular plane stress elements, and the kinematic boundary conditions are as follows: vertical displacement along edge FA is fixed, the other edges are free to move. The structure is subjected to uniaxial distributed load p, which linearly increases in time from zero to the critical value p_{cr} (Fig. 3.16).

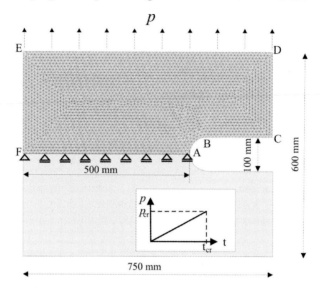

Fig. 3.16. Geometry, FE mesh and loading of the plane-stress structure

The structure is made of a high strength concrete. The material data that describe the basic local Murakami–Kamiya model are taken from Murakami and Kamiya [162], see (3.31). For the additional data, characteristic length of the non-local continuum (3.38) l=7.5 [mm], and critical fracture energy $G_c = 0.09$ [N/mm] (3.51) were adopted.

Crack propagation

The resulting crack propagation, together with the corresponding von Mises stress redistribution, is shown in Fig. 3.17. The macrocrack initiates in the element placed in the notch (Fig. 3.17a), where the stress concentration is observed. At the first stage of the cracking process it propagates along the axis of symmetry (Fig. 3.17b), but soon the direction of cracking changes, to lead finally to the failure mechanism shown in Fig. 3.17d.

a)

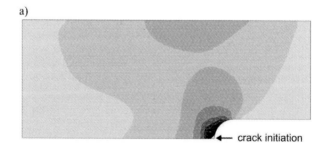

← crack initiation

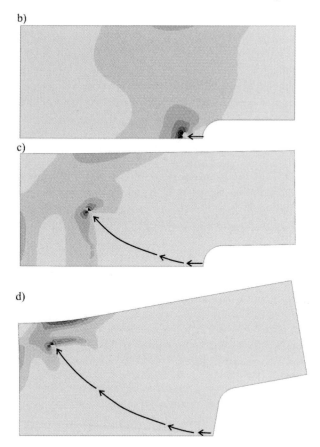

Fig. 3.17. Chosen stages of crack propagation together with von Mises stress redistribution, visualized on deformed mesh

3.2 Constitutive Models of Elasto-Plastic-Damage Materials

3.2.1 The Modified Hayakawa–Murakami Constitutive Model of Thermo-Elasto-Plastic-Damage Materials

State potential and state equations

Basic state and evolution equations for elasto-plastic-damage material of moderate ductility used in this chapter are originated in Hayakawa and Murakami [98], who implemented the total formulation. Extension of this model to the incremental formulation is due to Bielski *et al.* [22].

In contrast to the elastic-damage material model described in Sect. 3.1, the following material model allows for two dissipation mechanisms: plasticity and damage instead of one, by use of the concept of *weak dissipation coupling* (cf. 2.2.2). Moreover, to make the experimental verification straightforward, i.e. to have damage conjugate forces expressed in stresses instead of strains (cf. 3.1.1), the *Gibbs thermodynamic potential* is applied instead of the Helmholtz one (cf. 2.2.3) as the *state potential*.

$$\Gamma\left(\boldsymbol{\sigma},r,\mathbf{D},\beta\right)=\boldsymbol{\sigma}:\boldsymbol{\varepsilon}^{\mathrm{e}}-\Psi\left(\boldsymbol{\varepsilon}^{\mathrm{e}},r,\mathbf{D},\beta\right) \tag{3.52}$$

where $\Psi\left(\boldsymbol{\varepsilon}^{\mathrm{e}},r,\mathbf{D},\beta\right)$ is the Helmholtz free energy per unit mass, r is the scalar plastic isotropic hardening variable, $\mathbf{D}$ is the second-rank damage tensor, and β is an additional scalar variable to describe damage hardening. The Gibbs potential is composed of three terms: elastic, plastic and damage (2.21).

$$\Gamma\left(\boldsymbol{\sigma},r,\mathbf{D},\beta\right)=\Gamma^{\mathrm{e}}\left(\boldsymbol{\sigma},\mathbf{D}\right)+\Gamma^{\mathrm{p}}\left(r\right)+\Gamma^{\mathrm{d}}\left(\beta\right) \tag{3.53}$$

where only isotropic hardening is admitted, instead of both kinematic and isotropic hardening when a more general formulation is used (Abu Al-Rub and Voyiadjis [6]). Γ^{e} is the complementary energy due to the elastic deformation, which is assumed to be quadratic with respect to $\boldsymbol{\sigma}$ and linear with respect to $\mathbf{D}$. Additionally, the *unilateral damage response* in tension or compression is accounted for by a *modified stress tensor* $\boldsymbol{\sigma}^*$ defined in principal axes in an analogous fashion to (3.2):

$$\sigma_I^* = \langle\sigma_I\rangle - \zeta\langle-\sigma_I\rangle \equiv k\left(\sigma_I\right)\sigma_I \tag{3.54}$$
$$k\left(\sigma_I\right) = H\left(\sigma_I\right)+\zeta H\left(-\sigma_I\right) \quad (I=1,2,3,\text{no sum.})$$

where $\zeta\in\langle0,1\rangle$ is an additional *unilateral damage parameter*. Similar to Sect. 3.1.1, for $\zeta=1: \boldsymbol{\sigma}^*\equiv\boldsymbol{\sigma}$ (no unilateral response), whereas for $\zeta=0:$ $\boldsymbol{\sigma}^*\equiv\boldsymbol{\sigma}^+$ (Sect. 2.2.4).

Eventually, Γ^{e} is expressed as a linear combination of the basic six invariants of the tensors $\boldsymbol{\sigma}$ and $\mathbf{D}$, instead of the ten basic invariants in a general case (Spencer [213], Rymarz, [191])

$$\Gamma^{\mathrm{e}}\left(\boldsymbol{\sigma},\mathbf{D}\right)=-\frac{\nu}{2E}\left(\mathrm{Tr}\boldsymbol{\sigma}\right)^2+\frac{1+\nu}{2E}\mathrm{Tr}\left(\boldsymbol{\sigma}\cdot\boldsymbol{\sigma}\right)+\vartheta_1\mathrm{Tr}\mathbf{D}\left(\mathrm{Tr}\boldsymbol{\sigma}\right)^2 \tag{3.55}$$
$$+\vartheta_2\mathrm{Tr}\mathbf{D}\,\mathrm{Tr}\left(\boldsymbol{\sigma}^*\cdot\boldsymbol{\sigma}^*\right)+\vartheta_3\mathrm{Tr}\boldsymbol{\sigma}\,\mathrm{Tr}\left(\boldsymbol{\sigma}\cdot\mathbf{D}\right)+\vartheta_4\mathrm{Tr}\left(\boldsymbol{\sigma}^*\cdot\boldsymbol{\sigma}^*\cdot\mathbf{D}\right)$$

Note that $\boldsymbol{\sigma}^*$ is used only in the 4th and 6th terms of (3.55) to preserve the continuity requirement of the stress–strain response from crack opening to crack closure (Chaboche *et al.* [40])

With Γ^{e} described by (3.55) the following state equation of *anisotropic elasticity coupled with damage* is furnished by the use of conventional thermodynamic formalism (cf. Sect. 2.2.3)

$$\varepsilon^{\mathrm{e}} = \frac{\partial \Gamma^{\mathrm{e}}}{\partial \boldsymbol{\sigma}} = -\frac{\nu}{E}\left(\mathrm{Tr}\boldsymbol{\sigma}\right)\mathbf{1} + \frac{1+\nu}{E}\boldsymbol{\sigma} + 2\vartheta_1\left(\mathrm{Tr}\mathbf{D}\mathrm{Tr}\boldsymbol{\sigma}\right)\mathbf{1} + 2\vartheta_2\left(\mathrm{Tr}\mathbf{D}\right)\boldsymbol{\sigma}^* : \frac{\partial \boldsymbol{\sigma}^*}{\partial \boldsymbol{\sigma}}$$

$$+\vartheta_3\left[\mathrm{Tr}\left(\boldsymbol{\sigma}\cdot\mathbf{D}\right)\mathbf{1} + \left(\mathrm{Tr}\boldsymbol{\sigma}\right)\mathbf{D}\right] + \vartheta_4\left(\boldsymbol{\sigma}^*\cdot\mathbf{D} + \mathbf{D}\cdot\boldsymbol{\sigma}^*\right) : \frac{\partial \boldsymbol{\sigma}^*}{\partial \boldsymbol{\sigma}}$$

$$(3.56)$$

In the above, ν and E are the elastic constants for undamaged material, ϑ_1, ϑ_2, ϑ_3 and ϑ_4 are the material constants for the damage effect.

The part of the Gibbs potential $\Gamma^{\mathrm{p}}(r)$ represents the asymptotic isotropic hardening effect described by the cumulative plastic strain r.

$$\Gamma^{\mathrm{p}}\left(r\right) = R_{\infty}\left[r + \frac{1}{b}\exp\left(-br\right)\right]$$

$$(3.57)$$

where R_{∞} and b are plastic material constants.

$\Gamma^{\mathrm{d}}(\beta)$ is the part of the Gibbs potential related to damage. When a simple, linear relation between the internal damage hardening variable β and its conjugate force B is postulated, the following holds:

$$\Gamma^{\mathrm{d}}\left(\beta\right) = \frac{1}{2}K_{\mathrm{d}}\beta^2$$

$$(3.58)$$

where K_{d} is another damage material constant.

Hence, the *thermodynamic force-conjugates* $\mathbf{Y}$, R and B of *internal state variables* $\mathbf{D}$, r and β, are obtained as:

$$\mathbf{Y} \equiv \frac{\partial \Gamma^{\mathrm{e}}}{\partial \mathbf{D}} = \left[\vartheta_1\left(\mathrm{Tr}\boldsymbol{\sigma}\right)^2 + \vartheta_2\mathrm{Tr}\left(\boldsymbol{\sigma}^*\cdot\boldsymbol{\sigma}^*\right)\right]\mathbf{1} + \vartheta_3\left(\mathrm{Tr}\boldsymbol{\sigma}\right)\boldsymbol{\sigma} + \vartheta_4\left(\boldsymbol{\sigma}^*\cdot\boldsymbol{\sigma}^*\right)$$

$$R \equiv \frac{\partial \Gamma^{\mathrm{p}}}{\partial r} = R_{\infty}\left[1 - \exp\left(-br\right)\right]$$

$$B \equiv \frac{\partial \Gamma^{\mathrm{d}}}{\partial \beta} = K_{\mathrm{d}}\beta$$

$$(3.59)$$

The elastic-damage constitutive equation (3.56) represents the total formulation of the stress–elastic strain relation

$$\varepsilon^{\mathrm{e}} = {}^{\mathrm{s}}\tilde{\mathbf{C}}^{\mathrm{e}}\left(\mathbf{D}\right) : \boldsymbol{\sigma} \quad \text{or} \quad \{\varepsilon^{\mathrm{e}}\} = \left[{}^{\mathrm{s}}\tilde{\mathbf{C}}^{\mathrm{e}}\left(\mathbf{D}\right)\right]\{\boldsymbol{\sigma}\} \qquad (3.60)$$

where ${}^{\mathrm{s}}\tilde{\mathbf{C}}^{\mathrm{e}}\left(\mathbf{D}\right)$ or $\left[{}^{\mathrm{s}}\tilde{\mathbf{C}}^{\mathrm{e}}\left(\mathbf{D}\right)\right]$ represented the *effective elastic secant compliance tensor* or *matrix*, respectively, when the tensorial or the matrix–vector notation are used (cf. 2.34).

Dissipation potentials and evolution equations for plasticity and damage

The *dissipation potential* is assumed to be composed of two parts, plastic and damage (weak dissipation coupling), and the maximum dissipation principle holds

$$F\left(\boldsymbol{\sigma}, \mathbf{Y}, R, B; \ \mathbf{D}, r, \beta\right) = F^{\mathrm{p}}\left(\boldsymbol{\sigma}, R; \ \mathbf{D}\right) + F^{\mathrm{d}}\left(\mathbf{Y}, B; \ \mathbf{D}, r, \beta\right) \qquad (3.61)$$

Note that only isotropic plasticity and damage hardening are accounted for in the above formulation. Hence the thermodynamic force conjugates to the fluxes $\left\{\dot{\varepsilon}^{\mathrm{p}}, \dot{r}, \dot{\mathbf{D}}, \dot{\beta}\right\}$ are $\left\{\boldsymbol{\sigma}, R, \mathbf{Y}, B\right\}$. It introduces a limitation to the relatively simple loading conditions. If more complex loading history is considered, the kinematic hardening should also be included (Abu Al-Rub and Voyiadjis [6]). In what follows, having in mind that damage will, in general, influence the yield surface by the fourth-rank tensor function $\mathbf{M}(\mathbf{D})$, (Hayakawa and Murakami [98]), we assume

$$
\begin{aligned}
F^{\mathrm{p}}\left(\boldsymbol{\sigma}, R; \ \mathbf{D}\right) &= \sigma_{\mathrm{eq}} - \left(R_0 + R\right) = 0 \\
\sigma_{\mathrm{eq}} &= \sqrt{\tfrac{3}{2}\boldsymbol{\sigma}' : \mathbf{M}\left(\mathbf{D}\right) : \boldsymbol{\sigma}'}
\end{aligned}
\qquad (3.62)
$$

where R_0 denotes the initial yield stress and $\boldsymbol{\sigma}'$ the stress deviator. The fourth-rank tensor function $\mathbf{M}(\mathbf{D})$ linear in $\mathbf{D}$ is introduced to describe the effect of damage on the yield surface. The damage dissipation potential is defined in the space of force conjugate $\mathbf{Y}$ in the following analogous fashion:

$$
\begin{aligned}
F^{\mathrm{d}}\left(\mathbf{Y}, B; \ \mathbf{D}, r, \beta\right) &= Y_{\mathrm{eq}} + c^{\mathrm{r}}r\mathrm{Tr}\mathbf{D}\mathrm{Tr}\mathbf{Y} - \left(B_0 + B\right) = 0 \\
Y_{\mathrm{eq}} &= \sqrt{\tfrac{1}{2}\mathbf{Y} : \mathbf{L}\left(\mathbf{D}\right) : \mathbf{Y}}
\end{aligned}
\qquad (3.63)
$$

where the fourth-rank tensor function $\mathbf{L}\left(\mathbf{D}\right)$ is used to describe influence of the damage tensor on the damage surface.

Both tensor functions $\mathbf{M}\left(\mathbf{D}\right)$ and $\mathbf{L}\left(\mathbf{D}\right)$ in (3.62) and (3.63), respectively, are built in such a way that in pre-damage states the yield and damage functions are isotropic (cf. 2.29). Note, also, the additional 2nd term in (3.63) that couples the damage surface F^{d} and the cumulative plastic strain r. Constants R_0, B_0 and c^{p}, c^{r} that appear in (3.62), (3.63) and (2.29), are additional material parameters which govern the initial size and the effect of damage on both dissipation surfaces. Note, finally, that due to (3.59) Eqs. (3.62) and (3.63) describe two independent plastic and damage dissipation surfaces in the stress space. This means that, according to the model described in this section, damage and plastic dissipation can develop independently or simultaneously, such that the following material response may occur: pure elastic or elastic-damage or elastic-plastic or elastic-damage-plastic.

Using Eqs. (3.56) and (3.59) the *Clausius–Duhem inequality* (2.38) must hold with two constraints (3.62) and (3.63). Hence, by applying the *generalized normality rule* (Sect. 2.2.3), we arrive at the *plasticity* and *damage evolution equations* (2.40):

$$\dot{\varepsilon}^{\mathrm{p}} = \dot{\lambda}^{\mathrm{p}}\frac{\partial F^{\mathrm{p}}}{\partial \boldsymbol{\sigma}} = \dot{\lambda}^{\mathrm{p}}\frac{3}{2}\frac{\mathbf{M}(\mathbf{D}) : \boldsymbol{\sigma}}{\sigma_{\mathrm{eq}}} = \dot{\lambda}^{\mathrm{p}}\mathbf{m}$$

$$\dot{r} = \dot{\lambda}^{\mathrm{p}} \frac{\partial F^{\mathrm{p}}}{\partial(-R)} = \dot{\lambda}^{\mathrm{p}}$$

$$\mathbf{\dot{D}} = \dot{\lambda}^{\mathrm{d}} \frac{\partial F^{\mathrm{d}}}{\partial \mathbf{Y}} = \dot{\lambda}^{\mathrm{d}} \left[\frac{\mathbf{L} : \mathbf{Y}}{2Y_{\mathrm{eq}}} + c^{\mathrm{r}} r (\mathrm{Tr}\mathbf{D})\mathbf{1} \right] \qquad (3.64)$$

$$\dot{\beta} = \dot{\lambda}^{\mathrm{d}} \frac{\partial F^{\mathrm{d}}}{\partial(-B)} = \dot{\lambda}^{\mathrm{d}}$$

where new auxiliary second-rank tensor $\mathbf{m}$ was introduced for abbreviation.

The values of plastic multiplier $\dot{\lambda}^{\mathrm{p}}$ and damage multiplier $\dot{\lambda}^{\mathrm{d}}$ are obtained from the simultaneously solved consistency conditions

$$\dot{F}^{\mathrm{p}} \equiv \frac{\partial F^{\mathrm{p}}}{\partial \boldsymbol{\sigma}} : \dot{\boldsymbol{\sigma}} + \frac{\partial F^{\mathrm{p}}}{\partial R} \dot{R} + \frac{\partial F^{\mathrm{p}}}{\partial \mathbf{D}} : \mathbf{\dot{D}} = 0$$
$$\dot{F}^{\mathrm{d}} \equiv \frac{\partial F^{\mathrm{d}}}{\partial \mathbf{Y}} : \dot{\mathbf{Y}} + \frac{\partial F^{\mathrm{d}}}{\partial B} \dot{B} + \frac{\partial F^{\mathrm{d}}}{\partial \mathbf{D}} : \mathbf{\dot{D}} = 0 \qquad (3.65)$$

to finally arrive at

$$\dot{\lambda}^{\mathrm{d}} = \frac{\dfrac{\partial F^{\mathrm{d}}}{\partial \mathbf{Y}} : \dot{\mathbf{Y}}}{\dfrac{\mathrm{d}B}{\mathrm{d}\beta} - \dfrac{\partial F^{\mathrm{d}}}{\partial \mathbf{D}} : \dfrac{\partial F^{\mathrm{d}}}{\partial \mathbf{Y}}} \qquad \dot{\lambda}^{\mathrm{p}} = \frac{\dfrac{\partial F^{\mathrm{p}}}{\partial \boldsymbol{\sigma}} : \dot{\boldsymbol{\sigma}} + \left(\dfrac{\partial F^{\mathrm{p}}}{\partial \mathbf{D}} : \dfrac{\partial F^{\mathrm{d}}}{\partial \mathbf{Y}} \right) \dot{\lambda}^{\mathrm{d}}}{\dfrac{\mathrm{d}R}{\mathrm{d}r}} \qquad (3.66)$$

The thermodynamic force $\mathbf{Y}$ is expressed in terms of $\boldsymbol{\sigma}$ (3.59), such that its rate $\dot{\mathbf{Y}}$ may also be expressed by the rate of stress $\dot{\boldsymbol{\sigma}}$

$$\dot{\mathbf{Y}} = \frac{\partial \mathbf{Y}}{\partial \boldsymbol{\sigma}} : \dot{\boldsymbol{\sigma}} \qquad (3.67)$$

Hence, both plasticity and damage multipliers $\dot{\lambda}^{\mathrm{p}}$ and $\dot{\lambda}^{\mathrm{d}}$ are given in terms of $\dot{\boldsymbol{\sigma}}$ as

$$\dot{\lambda}^{\mathrm{d}} = \left[\frac{\dfrac{\partial F^{\mathrm{d}}}{\partial \mathbf{Y}} : \dfrac{\partial \mathbf{Y}}{\partial \boldsymbol{\sigma}}}{\dfrac{\mathrm{d}B}{\mathrm{d}\beta} - \dfrac{\partial F^{\mathrm{d}}}{\partial \mathbf{D}} : \dfrac{\partial F^{\mathrm{d}}}{\partial \mathbf{Y}}} \right] : \dot{\boldsymbol{\sigma}} = \boldsymbol{\Lambda}^{\mathrm{d}} : \dot{\boldsymbol{\sigma}} = \{\boldsymbol{\Lambda}^{\mathrm{d}}\} \{\dot{\boldsymbol{\sigma}}\} \qquad (3.68)$$

$$\dot{\lambda}^{\mathrm{p}} = \left[\frac{\dfrac{\partial F^{\mathrm{p}}}{\partial \boldsymbol{\sigma}}}{\dfrac{\mathrm{d}R}{\mathrm{d}r}} + \frac{\left(\dfrac{\partial F^{\mathrm{p}}}{\partial \mathbf{D}} : \dfrac{\partial F^{\mathrm{d}}}{\partial \mathbf{Y}} \right) \left(\dfrac{\partial F^{\mathrm{d}}}{\partial \mathbf{Y}} : \dfrac{\partial \mathbf{Y}}{\partial \boldsymbol{\sigma}} \right)}{\left(\dfrac{\mathrm{d}R}{\mathrm{d}r} \right) \left(\dfrac{\mathrm{d}B}{\mathrm{d}\beta} - \dfrac{\partial F^{\mathrm{d}}}{\partial \mathbf{D}} : \dfrac{\partial F^{\mathrm{d}}}{\partial \mathbf{Y}} \right)} \right] : \dot{\boldsymbol{\sigma}} = \boldsymbol{\Lambda}^{\mathrm{p}} : \dot{\boldsymbol{\sigma}} = \{\boldsymbol{\Lambda}^{\mathrm{p}}\} \{\dot{\boldsymbol{\sigma}}\}$$

$$(3.69)$$

where second-rank tensors $\boldsymbol{\Lambda}^{\mathrm{d}}$ and $\boldsymbol{\Lambda}^{\mathrm{p}}$ depend on the current material state variables $\boldsymbol{\sigma}$, $\mathbf{Y}$, $\mathbf{D}$, and r. In other words, in the stress space two coupled dissipation surfaces exist, the plastic surface $F^{\mathrm{p}} = 0$ and the damage surface $F^{\mathrm{d}} = 0$, which should be updated together on the subsequent loading steps $\Delta \boldsymbol{\sigma} = \boldsymbol{\sigma}^{n+1} - \boldsymbol{\sigma}^n$ (Fig. 2.1). It happens, however, that one of two multipliers $\dot{\lambda}^{\mathrm{p}}$ or $\dot{\lambda}^{\mathrm{d}}$ equals zero, then the system of equations (3.68, 3.69) reduces to $\dot{\lambda}^{\mathrm{p}} > 0$

or $\dot{\lambda}^{d} > 0$, for passive damage or plastic processes, respectively. On the other hand if both multipliers $\dot{\lambda}^{P}$ and $\dot{\lambda}^{d}$ equal zero a doubly-passive plastic and damage process takes place.

Incremental thermo-elasto-plastic-damage equations

In the above, when the Hayakawa–Murakami [98] concept was used, elastic-damage constitutive equations (3.56) are formulated in a *total form* (cf. 2.34)

$$\varepsilon^{e} = {}^{s}\tilde{\mathbf{C}}^{e}\left(\mathbf{D}\right) : \boldsymbol{\sigma} \tag{3.70}$$

where ${}^{s}\tilde{\mathbf{C}}^{e}$ denotes the *effective elastic secant compliance tensor*. Since the plasticity and damage constitutive equations (3.64) through (3.69) are given in the incremental form, in order to solve coupled elasto-plastic-damage process we also need the incremental form of the equation (3.70). The incremental form of the elasto-damage constitutive equation requires the *tangent elastic compliance matrix* $\tilde{\mathbf{C}}^{e}$, to be derived following rule (2.45), as described in Bielski *et al.* [22]. For numerical implementation to FEM code the vector/matrix notation of tensorial equations is more convenient. Hence, the elastic-damage equations can be written as

$$\{\varepsilon^{e}\} = \left[{}^{s}\tilde{\mathbf{C}}^{e}\right]\{\boldsymbol{\sigma}\} \tag{3.71}$$

or

$$\{\dot{\varepsilon}^{e}\} = \left[\tilde{\mathbf{C}}^{e}\right]\{\dot{\boldsymbol{\sigma}}\} \tag{3.72}$$

where the total or incremental form of the vector/matrix equations is used. Note that the local tangent elastic compliance matrix in (3.72) depends on stress, damage and its force conjugate as arguments $\tilde{\mathbf{C}}^{e}\left(\boldsymbol{\sigma}, \mathbf{D}, \mathbf{Y}\right)$

For numerical integration, at each integration point where the constitutive relations are to be integrated, it is assumed that the *state variables* $(\{\varepsilon\}, r, \{\mathbf{D}\}, \beta)$ together with the *conjugate forces* $(\{\boldsymbol{\sigma}\}, R, \{\mathbf{Y}\}, B)$ are known at the beginning of a loading step. In the frame of small strain theory the following strain rate decomposition holds:

$$\{\dot{\varepsilon}\} = \{\dot{\varepsilon}^{e}\} + \{\dot{\varepsilon}^{P}\} + \{\dot{\varepsilon}^{T}\} \tag{3.73}$$

whereas the stress rate $\{\dot{\boldsymbol{\sigma}}\}$ is sought for, along with conjugate pairs increase of $\dot{R}$ and $\dot{r}$, $\{\dot{\mathbf{Y}}\}$ and $\{\dot{\mathbf{D}}\}$, $\dot{B}$ and $\dot{\beta}$.

Taking into account (3.64) and (3.69) the plastic strain rate is also given in terms of the stress rate as follows:

$$\dot{\varepsilon}^{P} = \mathbf{m}\dot{\lambda}^{P} = \mathbf{m}\left(\mathbf{\Lambda}^{P} : \dot{\boldsymbol{\sigma}}\right) = \left(\mathbf{m}\mathbf{\Lambda}^{P}\right) : \dot{\boldsymbol{\sigma}} \tag{3.74}$$

or if vector/matrix notation is used

$$\{\dot{\varepsilon}^{\mathrm{P}}\} = [\{\mathbf{m}\,(\boldsymbol{\sigma}, \mathbf{D})\}\,\{\boldsymbol{\Lambda}^{\mathrm{P}}\,(\boldsymbol{\sigma}, \mathbf{D}, \mathbf{Y}, r)\}]\,\{\dot{\boldsymbol{\sigma}}\} = \left[\tilde{\mathbf{C}}^{\mathrm{P}}\,(\boldsymbol{\sigma}, \mathbf{D}, \mathbf{Y}, r)\right]\{\dot{\boldsymbol{\sigma}}\} \quad (3.75)$$

Combining (3.72), (3.73) and (3.75) we arrive at the *incremental thermo-elasto-plastic-damage constitutive equation*

$$\{\dot{\varepsilon}\} - \{\dot{\varepsilon}^{\mathrm{T}}\} = \left[\tilde{\mathbf{C}}^{\mathrm{e}}\right]\{\dot{\boldsymbol{\sigma}}\} + \left[\tilde{\mathbf{C}}^{\mathrm{P}}\right]\{\dot{\boldsymbol{\sigma}}\} = \left[\tilde{\mathbf{C}}^{\mathrm{epd}}\right]\{\dot{\boldsymbol{\sigma}}\} \quad (3.76)$$

where $\tilde{\mathbf{C}}^{\mathrm{epd}} = \tilde{\mathbf{C}}^{\mathrm{e}}\,(\boldsymbol{\sigma}, \mathbf{D}, \mathbf{Y}) + \tilde{\mathbf{C}}^{\mathrm{P}}\,(\boldsymbol{\sigma}, \mathbf{D}, \mathbf{Y}, r)$ denotes the local *tangent elasto-plastic-damage compliance matrix* that depends on state variables known from the starting equilibrium point. In the particular case of *material orthotropy*, the direction of which coincides with the damage principal directions, the matrix $\tilde{\mathbf{C}}$ saves the symmetry property of (2.47). However, in the case of general *anisotropy* a full representation of the compliance matrix $[\tilde{\mathbf{C}}]_{ij}$ has to be considered (3.14).

Matrix constitutive equations for plane stress $\sigma_{33} = \sigma_{23} = \sigma_{13} = 0$

In the particular case of *plane stress* the corresponding column vectors of stress, strain and damage are

$$\boldsymbol{\sigma} = \left\{\begin{array}{c} \sigma_{11} \\ \sigma_{22} \\ \sigma_{12} \end{array}\right\} \quad \boldsymbol{\varepsilon} = \left\{\begin{array}{c} \varepsilon_{11} \\ \varepsilon_{22} \\ \varepsilon_{33} \\ \varepsilon_{12} \end{array}\right\} \quad \mathbf{D} = \left\{\begin{array}{c} D_{11} \\ D_{22} \\ D_{33} \\ D_{12} \end{array}\right\} \quad \mathbf{Y} = \left\{\begin{array}{c} Y_{11} \\ Y_{22} \\ Y_{33} \\ Y_{12} \end{array}\right\} \quad (3.77)$$

whereas for the increments of damage conjugate force $\dot{\mathbf{Y}}$, Eq. (3.67) reads

$$\left\{\begin{array}{c} \dot{Y}_{11} \\ \dot{Y}_{22} \\ \dot{Y}_{33} \\ \dot{Y}_{12} \end{array}\right\} = \left[\begin{array}{ccc} \partial Y_{11}/\partial\sigma_{11} & \partial Y_{11}/\partial\sigma_{22} & \partial Y_{11}/\partial\sigma_{12} \\ \partial Y_{22}/\partial\sigma_{11} & \partial Y_{22}/\partial\sigma_{22} & \partial Y_{22}/\partial\sigma_{12} \\ \partial Y_{33}/\partial\sigma_{11} & \partial Y_{33}/\partial\sigma_{22} & \partial Y_{33}/\partial\sigma_{12} \\ \partial Y_{12}/\partial\sigma_{11} & \partial Y_{12}/\partial\sigma_{22} & \partial Y_{12}/\partial\sigma_{12} \end{array}\right] \left\{\begin{array}{c} \dot{\sigma}_{11} \\ \dot{\sigma}_{22} \\ \dot{\sigma}_{12} \end{array}\right\} \quad (3.78)$$

Finally, the *incremental constitutive equation* reduces to the plane-stress form

$$\left\{\begin{array}{c} \dot{\varepsilon}_{11} - \tilde{\alpha}_{11}\dot{\mathrm{T}} \\ \dot{\varepsilon}_{22} - \tilde{\alpha}_{22}\dot{\mathrm{T}} \\ \dot{\varepsilon}_{33} - \tilde{\alpha}_{33}\dot{\mathrm{T}} \\ \dot{\gamma}_{12} \end{array}\right\} = \left[\begin{array}{ccc} \tilde{C}_{11} & \tilde{C}_{12} & \tilde{C}_{16} \\ \tilde{C}_{12} & \tilde{C}_{22} & \tilde{C}_{26} \\ \tilde{C}_{13} & \tilde{C}_{23} & \tilde{C}_{36} \\ \tilde{C}_{16} & \tilde{C}_{26} & \tilde{C}_{66} \end{array}\right] \left\{\begin{array}{c} \dot{\sigma}_{11} \\ \dot{\sigma}_{22} \\ \dot{\sigma}_{12} \end{array}\right\} \quad (3.79)$$

When the rectangular (3×4) matrix (3.78) is contracted to the size (3×3), and thermal expansion coefficient has isotropic property, we have

$$\left\{\begin{array}{c} \dot{\varepsilon}_{11} - \alpha\dot{\mathrm{T}} \\ \dot{\varepsilon}_{22} - \alpha\dot{\mathrm{T}} \\ \dot{\gamma}_{12} \end{array}\right\} = \left[\begin{array}{ccc} \tilde{C}_{11} & \tilde{C}_{12} & \tilde{C}_{16} \\ & \tilde{C}_{22} & \tilde{C}_{26} \\ & & \tilde{C}_{66} \end{array}\right] \left\{\begin{array}{c} \dot{\sigma}_{11} \\ \dot{\sigma}_{22} \\ \dot{\sigma}_{12} \end{array}\right\} \quad (3.80)$$

The elements of tangent elasto-plastic-damage constitutive matrix $[\tilde{\mathbf{C}}]_{ij}$ are obtained on the basis of the general rule (2.45) through (2.46). Additionally, when the *unilateral damage* condition is considered, the unilateral fourth–rank transformation tensor $D_{ijkl} = \partial\sigma_{ij}^*/\partial\sigma_{kl}$ must be known explicitly

$$\mathrm{d}\sigma_{ij}^* = D_{ijkl}\mathrm{d}\sigma_{kl} \qquad D_{ijkl} = B_{ijkl} + \frac{\partial B_{ijpq}}{\partial\sigma_{kl}}\sigma_{pq} \qquad (3.81)$$

If plane stress is considered, the transformation from general to principal coordinates reduces to plane rotation by the angle Δ:

$$\Delta = \frac{1}{2}\arctan\frac{2\sigma_{12}}{\sigma_{11} - \sigma_{22}} \qquad (3.82)$$

and, hence, (3.81) reads

$$D_{ijkl} = B_{ijkl} + \frac{\partial B_{ijpq}}{\partial\Delta}\frac{\partial\Delta}{\partial\sigma_{kl}}\sigma_{pq} \qquad (3.83)$$

In what follows, to avoid cumbersome calculations, a simplified transformation rule is applied (Kuna-Ciskał and Skrzypek [137]), namely we assume

$$\begin{Bmatrix} \sigma_{11}^* \\ \sigma_{22}^* \\ \sigma_{12}^* \end{Bmatrix} = \begin{bmatrix} a & 0 & b \\ 0 & c & b \\ 0,5b & 0,5b & d \end{bmatrix} \begin{Bmatrix} \sigma_{11} \\ \sigma_{22} \\ \sigma_{12} \end{Bmatrix} \qquad (3.84)$$

where

$$\begin{aligned} a &= 0.5\,(k_1 + k_2) + 0.5\,(k_1 - k_2)\cos 2\Delta \\ b &= 0.5\,(k_1 - k_2)\sin 2\Delta \\ c &= 0.5\,(k_1 + k_2) - 0.5\,(k_1 - k_2)\cos 2\Delta \\ d &= 0.5\,(k_1 + k_2) \end{aligned} \qquad (3.85)$$

whereas $k_I = k(\sigma_I)$ denote coefficients in the stress modification rule (3.54)

$$\boldsymbol{\sigma}^* = \begin{Bmatrix} k_1\sigma_1 & & \\ & k_2\sigma_2 & \\ & & k_3\sigma_3 \end{Bmatrix} \qquad (3.86)$$

and take the value 1 or ζ depending on the positive or the negative principal stress sign at the point considered. In the particular case of lack of unilateral damage, $k_1 = k_2 = k_3 = 1$, such that the identical mapping holds (3.84). On the other hand, if $\Delta = 0$, we have

$$\sigma_{11}^* = k_1\sigma_{11} \qquad \sigma_{22}^* = k_2\sigma_{22} \qquad (3.87)$$

Finally, the *effective secant elastic compliance matrix* ${}^{\mathrm{s}}\tilde{\mathbf{C}}^{\mathrm{e}}$ (3.71) is furnished as follows (Bielski *et al.* [22]):

$$^s\tilde{C}^e_{11} = \frac{1}{E} + 2\vartheta_1 \text{Tr}\mathbf{D} + 2\vartheta_2 \text{Tr}\mathbf{D}(a^2 + 0.5b^2) + 2\vartheta_3 D_{11}$$

$$+\vartheta_4[D_{11}(2a^2 + 0.5b^2) + 0.5D_{22}b^2 + 2abD_{12}]$$

$$^s\tilde{C}^e_{12} = -\frac{\nu}{E} + 2\vartheta_1 \text{Tr}\mathbf{D} + \vartheta_2 b^2 \text{Tr}\mathbf{D} + \vartheta_3(D_{11} + D_{22})$$

$$+\vartheta_4[0.5b^2(D_{11} + D_{22}) + b(a + c)D_{12}]$$

$$^s\tilde{C}^e_{16} = 2b(a + d)\vartheta_2 \text{Tr}\mathbf{D} + 2\vartheta_3 D_{12}$$

$$+\vartheta_4[D_{11}b(2a + d) + bdD_{22} + 2D_{12}(b^2 + ad)]$$

$$^s\tilde{C}^e_{22} = \frac{1}{E} + 2\vartheta_1 \text{Tr}\mathbf{D} + 2\vartheta_2 \text{Tr}\mathbf{D}(c^2 + 0.5b^2) + 2\vartheta_3 D_{22}$$

$$+\vartheta_4[0.5D_{11}b^2 + D_{22}(2c^2 + 0.5b^2) + 2bcD_{12}]$$

$$^s\tilde{C}^e_{26} = 2b(c + d)\vartheta_2 \text{Tr}\mathbf{D} + 2\vartheta_3 D_{12} \tag{3.88}$$

$$+\vartheta_4[bdD_{11} + D_{22}b(2c + d) + 2D_{12}(b^2 + cd)]$$

$$^s\tilde{C}^e_{66} = 2\{\frac{1 + \nu}{E} + 2\vartheta_2 \text{Tr}\mathbf{D}(1 + b^2 + d^2)$$

$$+\vartheta_4[(b^2 + d^2)(D_{11} + D_{22}) + 4bdD_{12}]\}$$

Hence, in the *plane stress* condition the general *total form* of the elastic-damage matrix–vector constitutive equations (3.71) has the form

$$\left\{ \begin{array}{c} \varepsilon^e_{11} \\ \varepsilon^e_{22} \\ \gamma^e_{12} \end{array} \right\} = \left[\begin{array}{ccc} ^s\tilde{C}^e_{11} & ^s\tilde{C}^e_{12} & ^s\tilde{C}^e_{16} \\ & ^s\tilde{C}^e_{22} & ^s\tilde{C}^e_{26} \\ & & ^s\tilde{C}^e_{66} \end{array} \right] \left\{ \begin{array}{c} \sigma_{11} \\ \sigma_{22} \\ \sigma_{12} \end{array} \right\} \tag{3.89}$$

with the secant elastic compliance matrix elements $^s\tilde{C}^e_{ij}$ $(\mathbf{D})$ given by (3.88). The corresponding *incremental form* of the elastic-damage matrix–vector constitutive equations (3.72) is furnished in such a way that the new, *tangent elastic-damage compliance matrix* $\tilde{C}^e_{ij}(\boldsymbol{\sigma}, \mathbf{D}, \mathbf{Y})$ is defined according to the general rule (2.45) applied to the plain stress condition to arrive at (Bielski et al. [22])

$$\left\{ \begin{array}{c} \dot{\varepsilon}^e_{11} \\ \dot{\varepsilon}^e_{22} \\ \dot{\gamma}^e_{12} \end{array} \right\} = \left[\begin{array}{ccc} \tilde{C}^e_{11} & \tilde{C}^e_{12} & \tilde{C}^e_{16} \\ & \tilde{C}^e_{22} & \tilde{C}^e_{26} \\ & & \tilde{C}^e_{66} \end{array} \right] \left\{ \begin{array}{c} \dot{\sigma}_{11} \\ \dot{\sigma}_{22} \\ \dot{\sigma}_{12} \end{array} \right\} \tag{3.90}$$

The appropriate vector–matrix incremental plastic-damage equation (3.74) in the case of plane stress reduces to the following equation:

$$\left\{ \begin{array}{c} \dot{\varepsilon}^{\mathrm{p}}_{11} \\ \dot{\varepsilon}^{\mathrm{p}}_{22} \\ \dot{\gamma}^{\mathrm{p}}_{12} \end{array} \right\} = \left[\begin{array}{ccc} \tilde{C}^{\mathrm{p}}_{11} & \tilde{C}^{\mathrm{p}}_{12} & \tilde{C}^{\mathrm{p}}_{16} \\ & \tilde{C}^{\mathrm{p}}_{22} & \tilde{C}^{\mathrm{p}}_{26} \\ & & \tilde{C}^{\mathrm{p}}_{66} \end{array} \right] \left\{ \begin{array}{c} \dot{\sigma}_{11} \\ \dot{\sigma}_{22} \\ \dot{\sigma}_{12} \end{array} \right\} \qquad (3.91)$$

with the effective symmetric plastic-damage compliance matrix $\tilde{C}^{\mathrm{p}}_{ij}(\boldsymbol{\sigma}, \mathbf{D}, \mathbf{Y}, r)$ derived according to the formulae (3.75) and (3.64)

$$\left[\tilde{\mathbf{C}}^{\mathrm{p}} \right] = \left[\begin{array}{ccc} \tilde{C}^{\mathrm{p}}_{11} & \tilde{C}^{\mathrm{p}}_{12} & \tilde{C}^{\mathrm{p}}_{16} \\ & \tilde{C}^{\mathrm{p}}_{22} & \tilde{C}^{\mathrm{p}}_{26} \\ & & \tilde{C}^{\mathrm{p}}_{66} \end{array} \right] \qquad (3.92)$$

Eventually, with (3.76) taken into account, we arrive at (3.79):

$$\left\{ \begin{array}{c} \dot{\varepsilon}_{11} - \alpha_{11}\dot{T} \\ \dot{\varepsilon}_{22} - \alpha_{22}\dot{T} \\ \dot{\gamma}_{12} \end{array} \right\} = \left[\begin{array}{ccc} \tilde{C}^{\mathrm{e}}_{11} + \tilde{C}^{\mathrm{p}}_{11} & \tilde{C}^{\mathrm{e}}_{12} + \tilde{C}^{\mathrm{p}}_{12} & \tilde{C}^{\mathrm{e}}_{16} + \tilde{C}^{\mathrm{p}}_{16} \\ & \tilde{C}^{\mathrm{e}}_{22} + \tilde{C}^{\mathrm{p}}_{22} & \tilde{C}^{\mathrm{e}}_{26} + \tilde{C}^{\mathrm{p}}_{26} \\ & & \tilde{C}^{\mathrm{e}}_{66} + \tilde{C}^{\mathrm{p}}_{66} \end{array} \right] \left\{ \begin{array}{c} \dot{\sigma}_{11} \\ \dot{\sigma}_{22} \\ \dot{\sigma}_{12} \end{array} \right\} \qquad (3.93)$$

3.2.2 Application to Monotonic and Cyclic Uniaxial Loading Conditions

Computational algorithm

The iteration of global equilibrium of a system is performed by ABAQUS using the *Newton–Raphson method*. The method applied is based on the explicit forward Euler scheme, where the derivatives (stiffness or compliance) are known at the starting point on the loading path and kept constant along the increment. The procedure is not unconditionally stable, such that application of it is restricted to relatively simple loading history and sufficiently small incremental steps.

The particular forms of both elastic-damage (3.90) and plastic-damage (3.91) matrices depend not only on the state variables and conjugate forces, but also on whether the process on the loading step is active or passive for both damage and plasticity (see also Tab. 2.2). In general, in the case of *weak dissipation coupling* theory used in the model considered, two dissipation surfaces: *Plastic Limit Surface* (PLS) and *Damage Limit Surface* (DLS) are prescribed in the stress space, as shown schematically in Fig. 3.18.

The predictor–corrector approach was implemented using ABAQUS in Bielski *et al.* [22]. At each iteration step it starts with a doubly passive predictor, which means that matrix $\tilde{\mathbf{C}}^{\mathrm{e}}$ is determined for $\dot{\lambda}^{\mathrm{d}} = 0$ and matrix $\tilde{\mathbf{C}}^{\mathrm{p}}$ is zero because $\dot{\lambda}^{\mathrm{p}} = 0$, i.e. we assume there are no plastic strain increments and the elastic ones do not involve damage evolution. Afterwards, the stress increments are calculated, then $\left\{ \dot{\mathbf{Y}} \right\}$ (3.67) is found, and before the conjugate forces $\{\boldsymbol{\sigma}\}$ and $\{\mathbf{Y}\}$ are updated, checking is performed. In general, four cases are possible at the end of the increment:

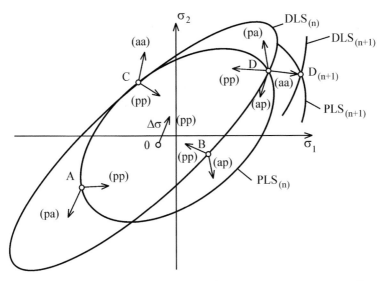

Fig. 3.18. Schematic illustration of damage/plastic, active or passive dissipation processes

1. None of the limit surfaces is exceeded (*doubly passive process*)

$$F^{\mathrm{d}}\left(\mathbf{Y}+\dot{\mathbf{Y}}, B, \mathbf{D}, r\right) \leq 0 \qquad \wedge \qquad F^{\mathrm{p}}\left(\boldsymbol{\sigma}+\dot{\boldsymbol{\sigma}}, R, \mathbf{D}\right) \leq 0 \qquad (3.94)$$

such that elastic increment is accepted, and the conjugate forces are updated:
$\boldsymbol{\sigma} \longleftarrow \boldsymbol{\sigma}+\dot{\boldsymbol{\sigma}},\ \mathbf{Y} \longleftarrow \mathbf{Y}\left(\boldsymbol{\sigma}\right)$

2. Only the damage surface is exceeded (*damage active process*)

$$F^{\mathrm{d}}\left(\mathbf{Y}+\dot{\mathbf{Y}}, B, \mathbf{D}, r\right) > 0 \qquad \wedge \qquad F^{\mathrm{p}}\left(\boldsymbol{\sigma}+\dot{\boldsymbol{\sigma}}, R, \mathbf{D}\right) \leq 0 \qquad (3.95)$$

hence the matrix $\tilde{\mathbf{C}}^{\mathrm{e}}$ is rebuilt following elastic-damage evolution ($\dot{\lambda}^{\mathrm{d}} > 0$, $\dot{\lambda}^{\mathrm{p}} = 0$), a new solution for stress increment $\dot{\boldsymbol{\sigma}}$ is obtained, $\dot{\lambda}^{\mathrm{d}}$, $\dot{\mathbf{Y}}, \dot{\mathbf{D}}$ and $\dot{\beta}$ are determined, and checking is done again for the plasticity limit: if

$$F^{\mathrm{p}}\left(\boldsymbol{\sigma}+\dot{\boldsymbol{\sigma}}, R, \mathbf{D}+\dot{\mathbf{D}}\right) \leq 0 \qquad (3.96)$$

the increment is accepted, the state variables $\mathbf{D}$ and β and the conjugate forces $\boldsymbol{\sigma}$, $\mathbf{Y}$ and B are updated: $\mathbf{D} \longleftarrow \mathbf{D}+\dot{\mathbf{D}}, \quad \beta \longleftarrow \beta+\dot{\beta}, \quad \boldsymbol{\sigma} \longleftarrow \boldsymbol{\sigma}+\dot{\boldsymbol{\sigma}},$ $\mathbf{Y} \longleftarrow \mathbf{Y}\left(\boldsymbol{\sigma}\right)$; else go to point (4).

3. Only the plastic surface is exceeded (*plastic active process*)

$$F^{\mathrm{d}}\left(\mathbf{Y}+\dot{\mathbf{Y}}, B, \mathbf{D}, r\right) \leq 0 \qquad \wedge \qquad F^{\mathrm{p}}\left(\boldsymbol{\sigma}+\dot{\boldsymbol{\sigma}}, R, \mathbf{D}\right) > 0 \qquad (3.97)$$

hence the matrix $\tilde{\mathbf{C}}^{\mathrm{p}}$ is rebuilt following plastic strain evolution ($\dot{\lambda}^{\mathrm{p}} > 0$, $\dot{\lambda}^{\mathrm{d}} = 0$), a new solution for stress increment $\dot{\boldsymbol{\sigma}}$ is obtained, $\dot{\lambda}^{\mathrm{p}}$, $\dot{r}$ and $\dot{\mathbf{Y}}$ are determined, and checking is done again for the damage limit: if

$$F^{\mathrm{d}}\left(\mathbf{Y} + \dot{\mathbf{Y}}, B, \mathbf{D}, r + \dot{r}\right) \leq 0 \tag{3.98}$$

the increment is accepted, the state variable r and the conjugate forces $\boldsymbol{\sigma}$, $\mathbf{Y}$ and R are updated: $r \longleftarrow r + \dot{r}$, $\boldsymbol{\sigma} \longleftarrow \boldsymbol{\sigma} + \dot{\boldsymbol{\sigma}}$, $R \longleftarrow R(r)$, $\mathbf{Y} \longleftarrow \mathbf{Y}(\boldsymbol{\sigma})$; otherwise go to point (4).

4. Both damage and plastic surface are exceeded (*doubly damage and plastic active process*)

$$F^{\mathrm{d}}\left(\mathbf{Y} + \dot{\mathbf{Y}}, B, \mathbf{D}, r\right) > 0 \quad \wedge \quad F^{\mathrm{p}}\left(\boldsymbol{\sigma} + \dot{\boldsymbol{\sigma}}, R, \mathbf{D}\right) > 0 \tag{3.99}$$

hence, both matrices $\tilde{\mathbf{C}}^{\mathrm{e}}$ and $\tilde{\mathbf{C}}^{\mathrm{p}}$ are rebuilt following plastic and damage evolution ($\dot{\lambda}^{\mathrm{p}} > 0$, $\dot{\lambda}^{\mathrm{d}} > 0$), the state variables r, $\mathbf{D}$ and β and the conjugate forces $\boldsymbol{\sigma}$, $\mathbf{Y}$, R and B are updated: $r \longleftarrow r + \dot{r}$, $\mathbf{D} \longleftarrow \mathbf{D} + \dot{\mathbf{D}}$, $\beta \longleftarrow \beta + \dot{\beta}$, $\boldsymbol{\sigma} \longleftarrow \boldsymbol{\sigma} + \dot{\boldsymbol{\sigma}}$, $\mathbf{Y} \longleftarrow \mathbf{Y}(\boldsymbol{\sigma})$, $R \longleftarrow R + \dot{R}$, $B \longleftarrow B + \dot{B}$.

Eventually, after one of the four above conditions holds, the local stiffness (or compliance) matrix (treated as tangent at the beginning of the step), possibly corrected after predictor–corrector treatment, is accepted to establish the global stiffness of a system.

If the initial point lies inside both surfaces O, the stress increment is doubly passive (pp). At the point on the plastic surface but inside the current damage surface A, the stress increment may be either doubly passive (pp) or passive–active (pa). At the point on the damage surface but inside current plastic surface B, stress increment may be either doubly passive (pp) or active–passive (ap). At the point were both surfaces meet tangentially C, the doubly passive or doubly active stress increment may occur. Eventually, if the stress point lies on the intersection of both limit surfaces D, four situations are possible for stress increment: doubly passive (pp) or doubly active (aa) or active–passive (ap) or passive–active (pa).

Example of uniaxial, uniform stress state – unilateral effect

The elasto-plastic-damage model described in 3.2.1 was calibrated for spheroidized graphite cast iron FCD400 by Hayakawa and Murakami [98]. It will be used as a reference material for further simulations (Table 3.1).

A manufacturing treatment that consists in adding particulate graphite to the cast iron is used in order to improve ductility, such that two dissipative mechanisms may occur: damage growth and/or plastic flow, instead of pure damage in the case of non-improved cast iron. Both mechanisms, damage and plastic, cause nonlinearity on the stress–strain curve that result in strain

Table 3.1. Calibration of an elastic (moderate) plastic-damage material model for the spheriodized graphite cast iron FCD400 (after Hayakawa and Murakami [98])

E[GPa]	ν[-]	ϑ_1[MPa^{-1}]	ϑ_2[MPa^{-1}]	ϑ_3[MPa^{-1}]	ϑ_4[MPa^{-1}]	ζ[-]	K_d[MPa]
169	0.285	-3.95×10^{-7}	4.00×10^{-6}	-4.00×10^{-7}	2.50×10^{-6}	0.89	1.3

b[-]	R_0[MPa]	R_∞[MPa]	B_0[MPa]	c^P[-]	c^d[-]	c^r[-]	ρ[kgm^{-3}]
15	293.0	250.0	0.273	1.0	-15.0	50.0	7840

softening. Usually brittle elastic-damage growth (without plasticity) exhibits abrupt, avalanche nature. Including "additional" ductility results in decreasing damage growth rate and, as a consequence, improved structure lifetime with respect to failure. However, on the other hand, some plastic strains can appear in a structure that cause irreversible deformation. Following the paper by Hayakawa and Murakami [98], Fig. 3.19, shows micrographs of a sample of spheriodized graphite cast iron subjected to uniaxial tension at the initial (damage free) state, and at the critical damage state ($\sigma_{\mathrm{cr}} = 498$ MPa, $\varepsilon_r = 0.15$)

a)   b)

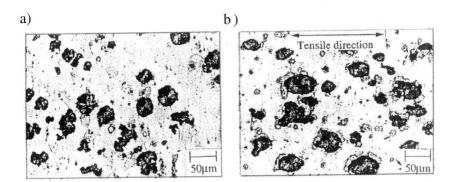

Fig. 3.19. Microstructure change in spheriodized graphite cast iron: (a) initial damage free state (b) deformation of the graphite particles accompanied by the damage growth at boundaries under tension (after Hayakawa and Murakami [98])

Size growth of the graphite particles and the shape deformation from a spherical (initial state) to an ellipsoidal (critical damage state) associated with damage evolution (decohesion), localized mainly at particle boundaries, is clearly observed. The figure explains also a mechanism for the damage-induced anisotropy, although in the virgin state a material behaves isotropically. Let us mention also that the constitutive model developed in 3.2.1 is based on the homogenization of physical properties of the two-phase (graphite/ferrite) material inside the Representative Volume Element (RVE). Changing the volume fraction of the graphite phase with respect to the ferritic matrix, as well as the size of graphite particles, the effective material properties of Table 3.1 may be changed as well, to allow for a mixed dissipation mechanism ranging from that of brittle to ductile (Table 4.11). Application of the constitutive model to other two-phase composite materials may also be considered, providing that a simplified description of moderate plastic deformation by the use of an isotropic plastic hardening model (3.62) is sufficient. In a more general case of advanced plastic response and/or cyclic loading conditions the extended formulae (2.30) can be used to define dissipation surfaces, where kinematic hardening is included.

In order to verify model capability to capture the following responses, elastic, elastic–damage or elastic-plastic-damage, simple uniaxial monotonic and cyclic loading conditions are used for a uniform stress state in a plane rectangular shield used as the structural model. Strain control is used because, for a stress-controlled process together with the forward Euler integration scheme and lack of consistent linearization, divergence might be met at the point-level integration, especially for the material model with asymptotic hardening.

Comparison of the experimental data (Hayakawa and Murakami [98]) and numerical simulation by the present model in the case of uniaxial tension is shown in Fig. 3.20.

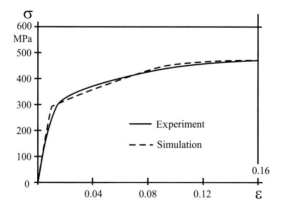

Fig. 3.20. Experimental vs. numerical stress–strain curve (Hayakawa and Murakami [98])

The initial doubly passive range (below the damage threshold $B_0 = 0.273$ MPa) is followed by the elastic-damage and concluded with the elastic-damage-plastic deformation (when the plastic limit $R_0 = 293$ MPa is exceeded). The effect of acquired anisotropy is confirmed in Fig. 3.21.

The diagonal components of the damage tensor: D_{11} and $D_{22} = D_{33}$, initially of value zero increase abruptly in the course of deformation, but the axial component D_{11} is higher than the transverse D_{22}. The evolution of plastic strains makes the damage rate slow down. With the progress of damage, deterioration of the elastic modulus occurs. This results in a stiffness drop, as shown in Fig. 3.22, initially rapid after the elastic-damage process is activated, then mitigated after plasticity is coupled with damage.

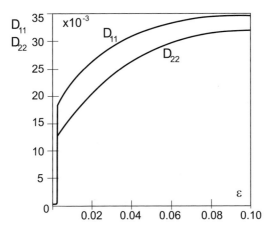

Fig. 3.21. Damage anisotropy and effect of plastic strains on damage rate

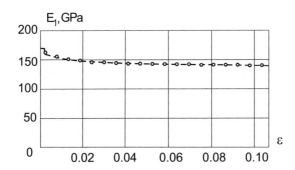

Fig. 3.22. Experimental vs. numerical stiffness drop due to elastic-damage and elastic-plastic-damage growth

In order to examine the model capability to capture the unilateral damage effect, a *cyclic strain-controlled* uniaxial loading *process* is simulated, as shown in Fig. 3.23a (Skrzypek *et al.* [208])

All phases of the deformation process are observed: elastic loading (e) → elastic-damage loading (ed) → elastic-damage-plastic loading (edp) → elastic unloading and reverse compressive loading (e) → reverse elastic-plastic loading (ep) → elastic unloading and secondary tensile loading (e) → secondary elastic-damage loading (ed) → secondary elastic-damage-plastic loading (edp), etc.

In Fig. 3.24 the single stress–strain loop (a) is presented together with the damage evolution on the cyclic loading steps (b) of Fig. 3.23.

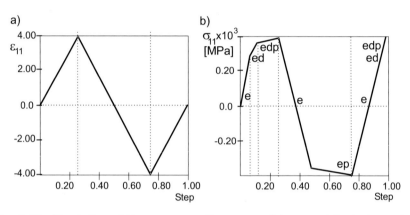

Fig. 3.23. Simulation of the reverse cyclic process: (a) symmetric strain control, (b) stress response (after Skrzypek et al. [208])

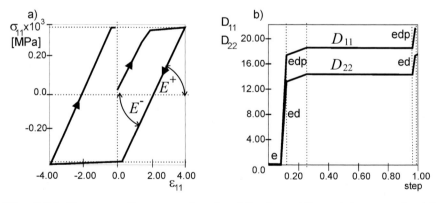

Fig. 3.24. Reverse cyclic process in spheroidized graphite cast iron sample: (a) stress–strain loop, (b) anisotropic damage evolution

On "damagely passive" phases (e) and (ep) the state of damage remains "frozen". Note also, that under the reverse compression a damage evolution is not observed, but only a reverse yielding occurs (ep). This is due to the fact that during the primary tensile phase the damage surface undergoes isotropic hardening associated with the elastic modulus drop $E_0 \rightarrow E^+$. When stress changes sign from tensile to compressive a partial elastic stiffness recovery occurs $E^- > E^+$, due to the unilateral effect included to the model (cf. 2.2.4). It is not clearly visible in Fig. 3.24a because the unilateral parameter ζ is, for the material considered, relatively high $\zeta = 0.89$ (cf. Table 3.1), such that the stiffness recovery on the reverse compression is hardly visible. In order to demonstrate a partial elastic stiffness recovery under compression the first component of the elastic stiffness matrix under plane condition $\tilde{C}_{11}^e = E/(1 - \nu^2)$ is checked over the whole loading cycle, as shown in Fig. 3.25.

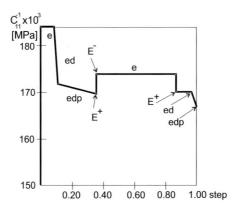

Fig. 3.25. Illustration of a partial elastic stiffness degradation/recovery over the reverse loading cycle

Spurious elastic modulus degradation is observed on the elastic-damage tension (ed) until a plastic tensile strain evolution (edp) mitigates the modulus drop to finally reach the value $E^+ = C_{11}^+/(1 - \nu)$ at the end of the "tensile phase". When changing next the stress from tensile to compressive, partial stiffness recovery of the material takes place $(E^- > E^+)$. Modulus E^- remains constant as long as the secondary tension begins, when the second stiffness jump from E^- to the last "remembered" value E^+ takes place. It is "frozen" again until, on a secondary tensile phase, further secondary elastic-damage process (ed) that followed by the elastic-damage-plastic response (edp) occurs.

3.2.3 Prediction of Damage and Plastic Evolution in the Plane notched FCD400 Cast Iron Specimen

In the previous Sect. 3.2.2 a uniform elastic-damage-plastic response to uniaxial tension/compression of a plane rectangular sample was analysed. In what follows, a non-uniform plane stress analysis of damage evolution and plastic strain growth in a double-notched sample subjected to uniformly applied tensile load is performed (Fig. 3.26)

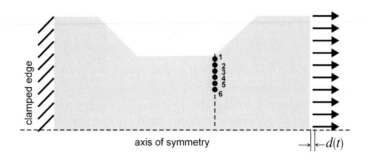

Fig. 3.26. Geometry, boundary and loading conditions of a double-notched plane sample

Non-uniform FE-mesh was used in Skrzypek *et al.* [208] in order to capture the effect of stress and damage localization around the notch. Control points 1 through 6 were chosen along the critical notch-through cross-section. With a monotonically increasing loading, controlled here by the edge displacement $d(t)$, evolution of variables: the equivalent stress σ_{eq}, predominant (axial) component of the damage tensor D_{11}, plastic hardening parameter r, and predominant component of the elasticity tensor E_{11} were numerically determined along critical cross-section (Fig. 3.27a–d).

Stress concentration at the notch is clearly visible (diagram a). Nonlinearities at stress–displacement diagrams appear after damage incubation (1st phase) and plastic incubation (2nd phase) periods at the points are exhausted. Damage evolution at the critical cross-section is presented in diagram b. The incubation time at the notch (point 1) is half that when compared to the interior (point 6). At the elastic-plastic phase abrupt damage growth is observed, as long as the plastic strain evolution at the point considered causes the damage rate to slow down (edp). With the plastic front propagating inwards on the sample, secondary slow down effects on the damage rate may also be observed (diagrams b and c). Note also that, for the data of Table 3.1, the plastic strain evolution is always preceded by damage growth at the point, and they interact each to the other in such a manner that plasticity mitigates damage growth at the considered point. On the other hand, the influence of both dissipative phenomena, plasticity and damage, at neighbouring points of the point considered, is strong. It is also clear when the stiffness degradation with damage is observed (diagram d).

In order to illustrate localization phenomena throughout the structure the variables of Fig. 3.27: σ_{eq}, D_{11}, r, and E_{11} are also presented in Fig. 3.28 in the form of bit-maps corresponding to the last loading step. Stress, damage

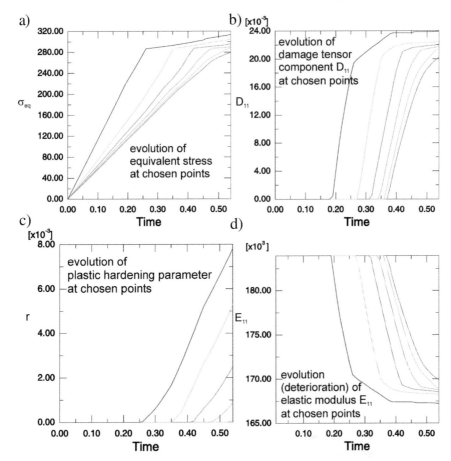

Fig. 3.27. Evolution of variables: σ_{eq}, D_{11}, r and E_{11} with edge displacement $d(t)$ along the critical cross-section of the sample (after Skrzypek *et al.* [208])

and plastic strain localization at the notches are clearly visible (a, b and c). It results in localized stiffness degradation propagating inwards from the notches in a similar fashion to the damaged zones (d versus b).

Damage and plastic strain localization around the notches, observed in Fig. 3.28b and c, cause a spurious mesh effect when the classical local constitutive model is used. To mitigate the mesh influence, nonlocal formulations of the constitutive equations should be used (Sect. 3.1.3). That is why the results presented above give the answer for a complex structural response, with two coupled dissipative mechanisms – plasticity and damage – included, in a qualitative sense only.

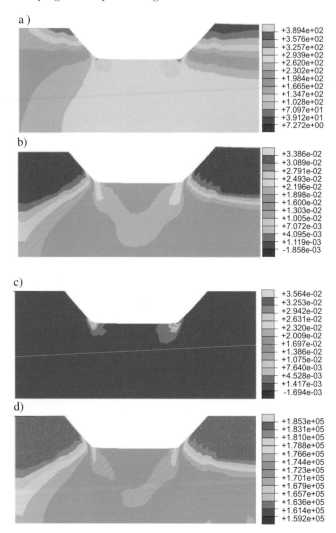

Fig. 3.28. Non-uniform spatial distribution of variables: σ_{eq}, D_{11}, r and $E_{11}(\mathbf{D})$ in the volume of a double-notched plane specimen at the last loading step (after Skrzypek *et al.* [208])

3.2.4 Low Cycle Fatigue of Specimen Made of 316L Stainless Steel

Experimental results

When a material is subjected to cyclic loading at high values of stress or strain, damage develops together with cyclic plastic strain. If the material is strain loaded, the damage induces a drop in the stress amplitude and a decrease of the elasticity modulus. Following Dufailly's experiment (Lemaitre [147]), a tension–compression test for AISI 316L stainless steel at constant amplitude of strain performed at room temperature, is shown in Fig. 3.29.

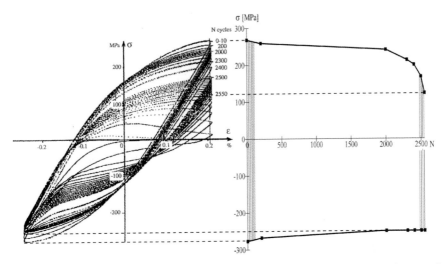

Fig. 3.29. Low cycle fatigue stress–strain behaviour for AISI 316L stainless steel Lemaitre [147]

The main difficulty involved in this is to avoid, at least partly, the localization by necking. The way to do this is to impose constant amplitude strain such that strain-hardening is saturated at the first cycle or after a few cycles. Detailed analysis of subsequent strain–stress loops reveals the existence of two softening processes corresponding to damage growth. During the first period ($0 < N < 2000$ cycles) damage growth manifested by transcrystalline microcracks is isotropic and exhibits a decreasing tendency. This results in an isotropic drop in both the stress amplitude and the elasticity modulus. With increasing number of cycles ($2000 < N < 2400$) another unilateral damage process, associated with cavities on the grain boundaries throughout the volume, is activated. The drop of stress amplitude and elasticity modulus is faster for tensile stress than for compressive stress. Finally, during the third stage directly preceding failure ($2400 < N < 2556$), the necking mechanism appears. It is manifested by formation of stress instability for tensile stress on the one hand, and the characteristic flex point on this part of the hysteresis, which corresponds to compressive stress, on the other hand.

Modelling strain–damage coupling

The existence of two separate damage mechanisms requires application of two independent scalar damage variables: the first isotropic damage variable D_s acts on the deviatoric stress and the second unilateral damage variable D_v

acts on the volumetric stress. A unilateral damage variable requires separate definitions of the effective stress, which allow one to distinguish the case of tension, when microcracks remain open, and the case of compression, when they are partly closed; effective stress takes therefore the following form:

$$\tilde{\sigma}^{\pm} = \frac{\sigma_s}{1 - D_s} + \frac{\sigma_v}{1 - D_v h} \qquad (3.100)$$

Smooth transition of the strain–stress relation between the elastic range and the nonlinear plastic range suggests application of the nonlinear approximations for the initial hysteresis loop. Hence, the Ylinen approximation (Ylinen [244]) is suggested since it gives the best fit to the experimental data:

$$\varepsilon = \frac{1}{E} \left[c\sigma - (1 - c)\sigma_0 \ln \left(1 - \frac{\sigma}{\sigma_0} \right) \right] \qquad (3.101)$$

c, σ_0 denote material constants ($0 \le c \le 1$).

In the case of material subjected to damage, the original Ylinen model has to be modified by replacement of the "nominal" stress σ by the effective stress $\tilde{\sigma}$ in (3.100). After differentiation with respect to stress, the damage-modified Ylinen approximation for loading cycles takes the following form:

$$\frac{d\sigma}{d\varepsilon} = \begin{cases} \tilde{E}^{\pm} \dfrac{\tilde{\sigma}_0^{\pm} - |\sigma|}{\tilde{\sigma}_0^{\pm} - c\,|\sigma|} & \text{loading} \\[2ex] \tilde{E}^{\pm} & \text{unloading} \end{cases} \qquad (3.102)$$

Inserting definitions of the effective stress (3.100), the effective damage modulus of elasticity and the effective asymptotic yield stress of Ylinen's model become functions of damage

$$\frac{\tilde{E}^{\pm}}{E} = \frac{\tilde{\sigma}_0^{\pm}}{\sigma_0} = \frac{(1 - D_s)(1 - D_v h)}{1 - \frac{2}{3}D_v h - \frac{1}{3}D_s} \qquad (3.103)$$

where crack closure parameter $h(\sigma)$ is given by (2.61).

The thermodynamically based damage evolution theory is due to Lemaitre and Chaboche [145, 147]. The main assumption is that in the case of low cycle fatigue, damage is related to the accumulated plastic strain. The potential of dissipation F is a sum of two parts, the first referring to Huber–Mises–Hencky yield condition f, where damage coupling acts only by the effective stress, and the other F_D associated with the kinetic law of damage evolution, being of the following form in the case of a uniaxial state of stress:

$$\begin{aligned} F &= f(\sigma, D_s, D_v) + F_D(Y, D_s, D_v) \\ f &= |\tilde{\sigma}| - R - \sigma_y \end{aligned} \qquad (3.104)$$

where the back stress is neglected in order to simplify mathematical formulae, whereas the isotropic hardening variable R is included into the Ylinen model.

Application of the classical formalism of associated plasticity leads to the uniaxial plastic strain expressed as

$$d\varepsilon^{P} = \frac{\partial F}{\partial \sigma} d\lambda = \frac{\partial F}{\partial \sigma_s} d\lambda = \text{sign}(\sigma_s) \frac{d\lambda}{1 - D_s} \quad (3.105)$$

and the accumulated plastic strain is given by

$$dp = |d\varepsilon^{P}| = \frac{d\lambda}{1 - D_s} \quad (3.106)$$

From the thermodynamical analysis point of view the damage conjugate variable is the strain energy density release rate Y. For the simplest expression of dissipation potential F_D it is based on kinetic damage of double scalar variables developed by Ladeveze [138] as follows:

$$F_D = \frac{Y_s^2}{2S_s(1 - D_s)} + \frac{Y_v^2}{2S_v(1 - D_v h)} \quad (3.107)$$

where S_s, S_v denote the deviatoric or volumetric part of a damage strength material parameter, respectively. The strain energy density release rate associated to the unilateral damage is given by

$$Y = Y_s + Y_v = \frac{\sigma^2}{3E(1 - D_s)^2} + \frac{\sigma^2}{6E(1 - D_v h)^2} \quad (3.108)$$

Since the damage rates are computed as follows:

$$dD_s = \frac{\partial F_D}{\partial Y_s} d\lambda \qquad dD_v = \frac{\partial F_D}{\partial Y_v} d\lambda \quad (3.109)$$

then taking advantage of (3.106), the final form of the kinetic damage law for the one-dimensional case is written as

$$\frac{dD_s}{dp} = \frac{\sigma^2}{3ES_s(1 - D_s)^2} g(p)$$
$$\frac{dD_v}{dp} = \frac{\sigma^2}{6ES_v(1 - D_v h)^2} \left(\frac{1 - D_s}{1 - D_v h}\right) H(p - p_D) \quad (3.110)$$

According to the derived evolution law the damage is related to plastic strain, proportional to accumulated plastic strain. Moreover, evolution of isotropic damage D_s is additionally controlled by the exponent function $g(p) = a\exp(-bp)$ in order to take into account its decreasing character. Also unilateral damage D_v is activated only when the accumulated plastic strain p reaches the damage threshold plastic strain p_D (Lemaitre [147]).

Approximate 3D description of damage evolution after strain localization

The proper description of the stage after strain localization requires analysis of the three-dimensional state of stress. Simultaneously, simplification of the numerical integration of the uniaxial model suggests application of a hybrid formulation, in which the stage before strain localization is modelled by uniaxial equations (3.102) and (3.110), whereas the stage after strain localization is modelled by an approximate solution for the 3D stress state, expressed as a uniform uniaxial stress beyond the localization zone, substituted into axial strain derived from 3D extension of the constitutive law. This allows one to perform the governing equations in simple form.

The simplest approximate 3D solution of the stress state accompanying axisymmetric strain localization is due to Davidenkov and Spiridonova [54]. Assuming material incompressibility, equality of hoop and radial logarithmic strain at the minimal cross-section, and curvature of an eigenstress trajectory at the minimal cross-section being in the form $1/\rho = r/(r_1\rho_1)$, the authors prove that the intensity of logarithmic strain is equal to the logarithmic axial strain (Fig. 3.30). The following approximate formulas were derived for stress

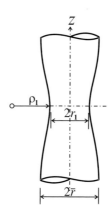

Fig. 3.30. Neck geometry

at the point belonging to the axis of symmetry at the minimal cross-section:

$$\overline{\sigma}_r = \overline{\sigma}_t = \overline{\sigma}_i \frac{r_1}{2\rho_1}, \quad \overline{\sigma}_z = \overline{\sigma}_i \left(1 + \frac{r_1}{2\rho_1}\right), \quad \overline{\sigma}_i = \left(\frac{\overline{r}}{r_1}\right)^2 \frac{\sigma}{1 + r_1/(4\rho_1)} \quad (3.111)$$

where $\overline{\sigma}_i = \overline{\sigma}_z - \overline{\sigma}_r$ denotes von Mises stress, whereas σ denotes a uniform uniaxial stress beyond the localization zone. The state of stress depends on two geometrical ratios $\overline{r}/r_1$ and r_1/ρ_1. The ratio of the radius of specimen $\overline{r}$ to the radius of neck r_1 results from an assumption of incompressibility and the definition of logarithmic strain $\overline{r}/r_1 = \sqrt{1 + \langle\overline{\varepsilon}\rangle}$, and it is approximated by unity for an applied amplitude of strain $\varepsilon_{\max} = 2.1 \times 10^{-3}$. On the contrary, the ratio of the radius of neck to the radius of curvature $r_1/\rho_1 \cong \bar{r}/\rho_1$ cannot be neglected and it is assumed in the form

$$\frac{\bar{r}}{\rho_1} = b\sqrt{\langle \varepsilon \rangle} \tag{3.112}$$

Strain under root appears in McAuley brackets since only its positive magnitude induces neck. Substitution of approximations derived above in (3.111) eventually gives approximate 3D formulae that describe axisymmetric stress state in a necked zone by nominal stress σ

$$\bar{\sigma}_r = \bar{\sigma}_t = \sigma \frac{0.5b\sqrt{\langle \bar{\varepsilon} \rangle}}{1 + 0.25b\sqrt{\langle \bar{\varepsilon} \rangle}}, \quad \bar{\sigma}_z = \sigma \frac{1 + 0.5b\sqrt{\langle \bar{\varepsilon} \rangle}}{1 + 0.25b\sqrt{\langle \bar{\varepsilon} \rangle}},$$
$$\bar{\sigma}_i = \frac{\sigma}{1 + 0.25b\sqrt{\langle \bar{\varepsilon} \rangle}} \tag{3.113}$$

In what follows an extension of the 1D damage modified Ylinen's constitutive law (3.101) for a 3D state of stress is used

$$\bar{\varepsilon}_{ij} = c\left[\frac{1}{2G}\frac{\bar{\sigma}_{ij}}{1 - D_s} + \frac{3}{K}\frac{\bar{\sigma}_v \delta_{ij}}{1 - D_v h}\right] - (1 - c)\frac{\sigma_0}{2G}\ln\left(1 - \frac{\bar{\sigma}_i}{\sigma_0}\right)\frac{3\bar{s}_{ij}}{2\sigma_0(1 - D_s)} \tag{3.114}$$

in which the first elastic term is compressible whereas the second inelastic term is incompressible, and $\bar{s}_{ij}$ denotes the deviatoric stress component. Hence, restricting (3.114) to axial strain we obtain an approximate formula for $\bar{\varepsilon}_z$ that accounts for the 3D axisymmetric stress at the neck

$$\bar{\varepsilon} = \bar{\varepsilon}_z = c\left[\frac{1}{2G}\frac{\bar{\sigma}_z}{1 - D_s} + \frac{3}{K}\frac{\bar{\sigma}_v}{1 - D_v h}\right] - (1 - c)\frac{\sigma_0}{2G}\ln\left(1 - \frac{\bar{\sigma}_z}{\sigma_0}\right)\frac{3\bar{s}_{ij}}{2\sigma_0(1 - D_s)} \tag{3.115}$$

Calculating $\bar{\sigma}_z$ and volumetric stress $\bar{\sigma}_v$ with Eqs. (3.113) taken into account, lead to

$$\bar{s}_z = \frac{2\bar{\sigma}_z - 2\bar{\sigma}_r}{3} = \frac{2\sigma}{3\left(1 + 0.25b\sqrt{\langle \bar{\varepsilon} \rangle}\right)}$$
$$\bar{\sigma}_v = \frac{2\bar{\sigma}_r + 2\bar{\sigma}_z}{3} = \frac{\sigma}{3}\frac{1 + 0.5b\sqrt{\langle \bar{\varepsilon} \rangle}}{1 + 0.25b\sqrt{\langle \bar{\varepsilon} \rangle}} \tag{3.116}$$

Next substitution of (3.116) together with σ_i in (3.113) to (3.115), when some simplifications are used give an approximate extension of 1D Ylinens law (3.102) to 3D axisymmetric stress state

$$\frac{d\sigma}{d\varepsilon} = \begin{cases} \hat{E}^\pm \dfrac{\hat{\sigma}_0^\pm - |\sigma|}{\hat{\sigma}_0^\pm - c|\sigma|} & \text{loading} \\ \hat{E}^\pm & \text{unloading} \end{cases} \tag{3.117}$$

where the new effective damage moduli of elasticity $\widehat{E}^{\pm}$ and the effective asymptotic yield stresses $\widehat{\sigma}_0^{\pm}$ of Ylinen's model (3.103) are defined by the following relations:

$$\frac{\widehat{E}^{\pm}}{\widetilde{E}^{\pm}} = \frac{\widehat{\sigma}_0^{\pm}}{\widetilde{\sigma}_0^{\pm}} = \frac{1 + 0.25b\sqrt{\langle \bar{\varepsilon} \rangle}}{1 + 0.5b\sqrt{\langle \bar{\varepsilon} \rangle}} \tag{3.118}$$

Note that, when a localization zone is not admitted, then following (3.112), which combines the neck geometry (radii $\bar{r}$ and ρ_1) with the approximate axial strain $\bar{\varepsilon} = \bar{\varepsilon}_z$ at the neck bottom, the formulae (3.118) yield the pure 1D formulae (3.103).

Extending now (3.104) to the 3D case yield condition takes the following form

$$F = f\left(\boldsymbol{\sigma}, D_{\mathrm{s}}, D_{\mathrm{v}}\right) + F_{\mathrm{D}}\left(Y, D_{\mathrm{s}}, D_{\mathrm{v}}\right)$$
$$f = \frac{\bar{\sigma}_{\mathrm{i}}}{1 - D_{\mathrm{s}}} - \sigma_{\mathrm{y}} \tag{3.119}$$

whereas approximate 3D plastic strain components are defined as follows

$$\mathrm{d}\bar{\varepsilon}_z^{\mathrm{p}} = \frac{\partial F}{\partial \sigma_z}\mathrm{d}\bar{\lambda} = \frac{\mathrm{d}\bar{\lambda}}{1 - D_{\mathrm{s}}} \qquad \mathrm{d}\bar{\varepsilon}_r^{\mathrm{p}} = \mathrm{d}\bar{\varepsilon}_t^{\mathrm{p}} = \frac{\partial F}{\partial \sigma_r}\mathrm{d}\bar{\lambda} = \frac{-\mathrm{d}\bar{\lambda}}{1 - D_{\mathrm{s}}} \tag{3.120}$$

whereas if the assumptions of the Davidenkov and Spiridionova approach are used, the accumulated plastic strain is given by the formula

$$\mathrm{d}\bar{p} = \sqrt{\frac{2}{3}\mathrm{d}\bar{\varepsilon}_{ij}^{\mathrm{p}}\mathrm{d}\bar{\varepsilon}_{ij}^{\mathrm{p}}} = \frac{\sqrt{2}\mathrm{d}\bar{\lambda}}{1 - D_{\mathrm{s}}} \tag{3.121}$$

The strain energy density release rate associated to the unilateral damage takes the form

$$Y_{\mathrm{s}} = \frac{\bar{\sigma}_{\mathrm{i}}^2}{2E\left(1 - D_{\mathrm{s}}\right)^2} = \frac{\sigma^2}{2E\left(1 - D_{\mathrm{s}}\right)^2}\frac{1}{\left(1 + 0.25b\sqrt{\langle \bar{\varepsilon} \rangle}\right)^2}$$

$$Y_{\mathrm{v}} = \frac{3\bar{\sigma}_{\mathrm{v}}^2}{2E\left(1 - D_{\mathrm{v}}h\right)^2} = \frac{\sigma^2\left(1 - D_{\mathrm{s}}\right)}{6E\left(1 - D_{\mathrm{v}}h\right)^3}\left[\frac{1 + 1.5b\sqrt{\langle \bar{\varepsilon} \rangle}}{1 + 0.25b\sqrt{\langle \bar{\varepsilon} \rangle}}\right]^2 H\left(\bar{p} - p_{\mathrm{D}}\right) \tag{3.122}$$

consequently the 3D extension of 1D kinetic damage law (3.110) is written as

$$\frac{\mathrm{d}D_{\mathrm{s}}}{\mathrm{d}\bar{p}} = \frac{\sigma^2}{2\sqrt{2}ES_{\mathrm{s}}\left(1 - D_{\mathrm{s}}\right)^2}\frac{1}{\left(1 + 0.25b\sqrt{\langle \bar{\varepsilon} \rangle}\right)^2}g\left(\bar{p}\right)$$

$$\frac{\mathrm{d}D_{\mathrm{v}}}{\mathrm{d}\bar{p}} = \frac{\sigma^2}{6\sqrt{2}ES_{\mathrm{v}}\left(1 - D_{\mathrm{v}}h\right)^2}\left[\frac{1 - D_{\mathrm{s}}}{1 - D_{\mathrm{v}}h}\right]\left[\frac{1 + 1.5b\sqrt{\langle \bar{\varepsilon} \rangle}}{1 + 0.25b\sqrt{\langle \bar{\varepsilon} \rangle}}\right]^2 H\left(\bar{p} - p_{\mathrm{D}}\right) \tag{3.123}$$

Eventually, in what follows, in order to account for both pre-necking and necking stages, we shall apply the 1D formulae (3.117) and (3.123) equivalent to the appropriate 3D stress state, respectively. This approximation enables one to simulate experimental results by Dufailly for a 1D reverse tension–compression test, where the stage of necking is taken into account. It will be shown that this concept allows prediction of the full range of low cycle fatique including instability due to necking.

Identification of model parameters

The magnitudes of material parameters related to elasticity and plasticity, namely E, ν and σ_y are cited by Lemaitre [147]. The material parameters of Ylinen's model are calibrated in the following way: the magnitude of c is set in order to close the first hysteresis loop, whereas the magnitude of σ_0 is established by comparison of the first hysteresis loop with that of a numerical simulation and minimization of the respective differences (cf. Fig. 3.31).

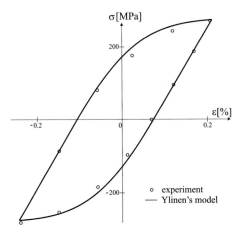

Fig. 3.31. Calibration of Ylinen's model parameters for the first hysteresis loop (Ganczarski and Barwacz [83])

Identification of the material parameters describing the kinetic law of damage evolution h_c, a, S_v, S_s, p_D is based on the procedure proposed by Lemaitre [147] in the case of isotropic damage and extended by Garion and Skoczeń [85] for the case of orthotropic damage. In particular, the critical magnitude of damage closure/opening parameter h_c generally depends on the material and loading. In practice, however, it is considered to be constant in order to simplify algebraic calculations. Lemaitre [147] recommends a value $h_c = 0.2$ as universal for a wide class of materials.

Complete material data of AISI 316L stainless steel at room temperature are presented in Table 3.2.

Table 3.2. Material data for AISI 316L stainless steel at room temperature (after Ganczarski and Barwacz [83])

a	b	c	E [GPa]	ν	h_c	p_D	S_v [MPa]	S_s [MPa]	σ_0 [MPa]	σ_y [MPa]
2.5	5.0	0.5773	200	0.3	0.2	3.2	0.151	0.25	280	260

Results

A system of three ordinary differential equations (3.102) and (3.110) for the stage preceding, or (3.117) and (3.123) for the stage after strain localization, is numerically integrated for strain from $\varepsilon_{\min} = -2.45 \times 10^{-3}$ to $\varepsilon_{\max} = 2.1 \times 10^{-3}$ using the `odeint.for` routine using fourth-order Runge–Kutta technique with adaptive stepsize control (Press *et al.* [183]).

Numerical simulation with sampling model of crack closure/opening effect excluded ($h = 1.0$), originally proposed by Lemaitre [147], is presented in Fig. 3.32. The model under consideration exhibits isotropic damage softening for tension and compression which disqualifies it from proper modelling of experimental results.

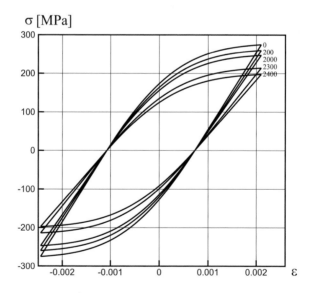

Fig. 3.32. Numerical simulation with sampling model of crack closure/opening effect excluded ($h = 1.0$) (Ganczarski and Barwacz [83])

Predictions based on a more advanced model of the classical crack closure/opening effect ($h = h_c$) give quantitatively good agreement with experimental data (Fig. 3.33). The model properly maps the unilateral nature of

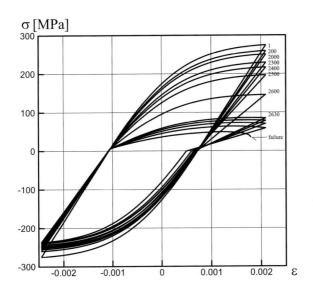

Fig. 3.33. Numerical simulation with model of classical crack closure/opening effect ($h = h_c$) (Ganczarski and Barwacz [83])

damage softening in the sense that dead centre ordinates of subsequent hysteresis loops coincide exactly with appropriate points on the experimental curves. However, the model exhibits $\partial\sigma/\partial\varepsilon$ discontinuity for $\sigma = 0$, leading to drastic disagreement with experiment for the stage after strain localization.

In contrast to the above models, numerical simulation with the model of continuous crack closure/opening effect included ($h(\sigma)$ according to (2.61)) exhibits not only quantitative but also qualitative proper correctness in comparison with experimental results (Fig. 3.34). The essential defect of the previous model, the existence of nonsmooth, bilinear material characteristic separating tensile and compression, manifested specially at the point of zero ordinate lying on tensile unloading to compression reloading branch of the hysteresis loop, is successfully eliminated. The main limitation of the model seems to be bad mapping of this part of hysteresis, which corresponds to compressive stress and consequently a lack of the characteristic point of zero curvature. Use of damage deactivation expressed in terms of total plastic strain (Chaboche [37]) instead of the presented stress formulation may improve the solution, however, it does not guarantee that the inflection point appears.

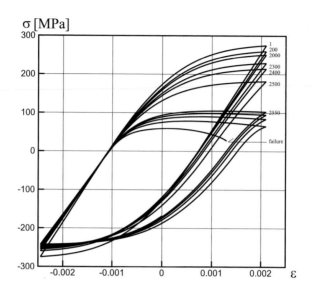

Fig. 3.34. Numerical simulation with model of continuous crack closure/opening effect included ($h(\sigma)$ according to (2.61)) (Ganczarski and Barwacz [83])

3.2.5 Influence of Unilateral Damage on the Yield Potential

Damage evolution equations presented in the previous section are derived on the basis of the kinetic law of damage evolution proposed by Lemaitre and Chaboche [145, 147]. The key point of this theory is the potential for dissipation taken as the sum of the yield potential, referring to the theory of associated plasticity, and a damage dissipation potential. In the simplest case of isotropic damage it is assumed that the kinetic coupling acts only by the effective stress deviator

$$\widetilde{s} = \frac{s}{1 - Dh} \tag{3.124}$$

and it corresponds to

$$F = \sqrt{\frac{3}{2}\widetilde{s} : \widetilde{s}} - \sigma_y + F_D\,(Y, D) \tag{3.125}$$

It is postulated that the dissipation potential F is a scalar continuous or partly continuous (having at most a finite set of corner points) and convex function.

The form of the yield potential essentially differs in the case of tension, when damage is active, from the case of compression, when damage remains inactive, and it is strictly associated with the magnitude of the damage de-activation parameter h in (3.124). As a consequence, in the case of the plane stress state and under the assumption that damage is active if only one of the stress components is positive ($\sigma_1 > 0$ or $\sigma_2 > 0$) or in other words $h = 1$ if $\mathrm{Tr}\,\langle\boldsymbol{\sigma}\rangle$ is positive, the yield potential takes following form:

$$\frac{\sigma_1^2 - \sigma_1\sigma_2 + \sigma_2^2}{\sigma_y^2} = \begin{cases} (1-D)^2 & \text{for } 1^{\text{st}},\ 2^{\text{nd}} \text{ and } 4^{\text{rth}} \text{ quarters} \\ 1 & \text{for } 3^{\text{rd}} \text{ quarter} \end{cases} \qquad (3.126)$$

and turns out to be nonsmooth and nonconvex (see Fig. 3.35). Namely, for any nonzero damage state (here $D = 0.6$) there exist two linear segments $A_1 A_2$ and $B_1 B_2$ corresponding to instantaneous damage deactivation that link together ellipses of second and third or third and fourth quarters, respectively. The above mentioned defect of the yield potential (3.126) may be successfully

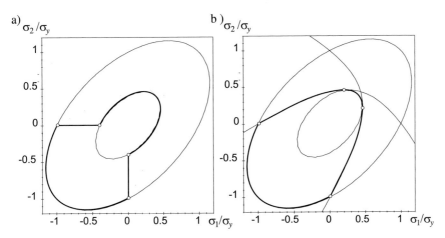

Fig. 3.35. Yield potential in the case of classical damage deactivation effect (a) and continuous damage effect (b) (Ganczarski and Cegielski [84])

removed by introducing continuous damage deactivation, when microcracks do not close instantaneously but gradually (2.68), and restricting considerations to the simplified case when $\chi(\boldsymbol{\sigma})$ in (2.69) depends only on positive value of the first invariant of stress tensor ($\alpha = 1$ and $\mathrm{Tr}(\boldsymbol{\sigma}) = \mathrm{Tr}\,\langle\boldsymbol{\sigma}\rangle$). Assuming further that microcracks are fully open under maximum tension $\sigma_b = 2(1-D)\sigma_y/\sqrt{3}$ (points A_1 and B_1) and close completely under compression ($h_c = 0$ at points A_2 and B_2) the yield potential is given by

$$\frac{\sigma_1^2 - \sigma_1\sigma_2 + \sigma_2^2}{\sigma_y^2} = \begin{cases} (1-D)^2 & \text{between points } A_1 \ \& \ B_1 \\ \left(1 - \dfrac{D}{1-D}\dfrac{\sqrt{3}}{2}\dfrac{\sigma_2}{\sigma_y}\right)^2 & \text{between points } A_1 \ \& \ A_2 \\ \left(1 - \dfrac{D}{1-D}\dfrac{\sqrt{3}}{2}\dfrac{\sigma_1}{\sigma_y}\right)^2 & \text{between points } B_1 \ \& \ B_2 \\ 1 & \text{for } 3^{\text{rd}} \text{ quarter between} \\ & \text{points } A_2 \ \& \ B_2 \end{cases} \qquad (3.127)$$

and is shown schematically in Fig. 3.36. Now yield potential is composed of two ellipses and two hyperbolas, becoming parabolas in the particular case $D = 2\sqrt{3}/(3 + 2\sqrt{3}) \cong 0.5358$, and its convexity and smoothness (except points A_2 and B_2) are recovered. Subsequent stages of damage affected yield potential versus experimental investigations after Litewka [151] are shown in Fig. 3.36.

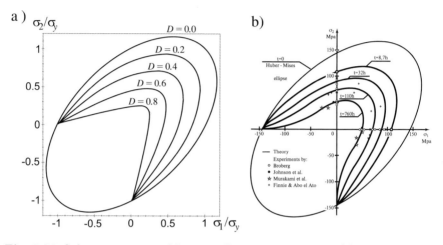

Fig. 3.36. Subsequent stages of damage affected yield potential (a) and experimental investigations after Litewka [151] (b) (Ganczarski and Cegielski [84])

4

Developing and Implementing Constitutive Models for Specific FGM Applications

4.1 Simulation Methodology for Properties Gradation in FGMs

In general, there are two approaches to modelling the material property gradation in FGMs, depending on the grade of coupling between the microscale and macroscale.

Uncoupled Approaches (UA) ignore the effect of microstructure gradation by employing specific spatial effective property variations that are either assumed or obtained by *local homogenization*, typically Reuss and Voigt methods, the Mori–Tanaka method, or the self consistent technique (for review see Mura [160]). The property gradations are then defined in the macroscopic analysis through mathematical functions or, in numerical analyses, at discrete points, with again two possibilities, a continuous variation or in a piecewise continuous manner in layered composites. Generally, linear, quadratic and cubic polynomials, exponential functions or piecewise homogeneous layer representations are used. The macroscopic approach is widely used, because of its simplicity, but this remains a simplification with no insight into the local microstructure. The local effect cannot always be neglected in the case of FGMs, especially in the case of high property gradients at the narrow interface layer, or for thin structures compared to the inclusions or defect size.

On the contrary, the *Coupled Approaches* (CA) take the effect of microstructure gradation into account and do not neglect the local–global interaction of the spatially variable inclusion phases. They are, however, time consuming and related to a particular distribution. Aboudi, Pindera and Arnold [2, 3, 5] developed *Higher Order Theory* for FGMs called HOTFGM, that explicitly couples the local (micro) and the global (macro) properties. In the so called efficient HOTFGM currently developed by Bansal and Pindera [13], the volume-averaged quantities used in the original HOTFGM approach are replaced by the surface-averaged quantities as the fundamental unknowns. The majority of papers deal with the *one-dimensional spatial gradation* of material properties of FGMs, Fig. 4.1, however in some cases the *two-dimensional gradation* needs to be used, but the appropriate fabrication techniques are rared available.

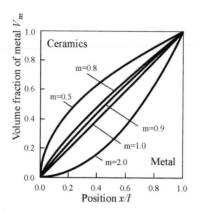

Fig. 4.1. The compositional profiles of unidirectional composites (after Fujimoto and Noda [77])

Hence, Aboudi, Pindera and Arnold [4] extended the one-dimensional approach by creating a two-dimensional framework HOTFGM–2D, to enable modelling of property composition graded in two directions.

Due to the fabrication techniques (e.g. in *Plasma Spray Thermal Barrier Coatings* PS–TBC deposition) the *graded material* should sometimes be considered as *orthotropic* rather than *isotropic*. Additionally, even if graded material is initially isotropic, *acquired anisotropy (orthotropy)* may result from damage and fracture growth (Skrzypek and Ganczarski [206]).

When the finite element method (FEM) based on *homogeneous elements* (HE) is used for FGMs the material properties vary in a piecewise continuous manner, from one element, to the other, but all integration points within an element have a common property value. The stiffness matrix of an element is given by the following integral:

$$\mathbf{k}^{(e)} = \int_{\Omega^{(e)}} \mathbf{B}^{(e)T} \mathbf{D}^{(e)} \mathbf{B}^{(e)} \mathrm{d}\Omega^{(e)} \tag{4.1}$$

where constitutive matrix $\mathbf{D}^{(e)}$ does not depend on coordinates, hence it is treated as constant at the level of the element. On the other hand, a too coarse mesh may lead to unrealistic stresses at the interface between the numerical layers. To overcome this difficulty a special *Graded Element* (GE) rather than *Homogeneous Element* (HE) has been introduced by Kim and Paulino [126] to discretize FGM properties. The material properties at Gauss quadrature points are interpolated there from the nodal material properties by the use of *isoparametric interpolation functions*, as shown schematically in Fig. 4.2 (Kim

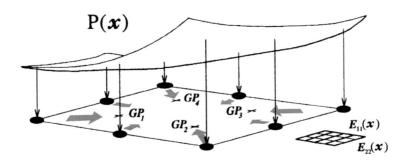

Fig. 4.2. Generalized isoparametric function formulation using GE (after Kim and Paulino [125])

and Paulino [125]). Contrary to the classical FEM formulation, the stiffness matrix of an element is expressed by the integral

$$\mathbf{k}^{(e)} = \int_{\Omega^{(e)}} \mathbf{B}^{(e)T} \mathbf{D}^{(e)}\left(\mathbf{x}\right) \mathbf{B}^{(e)} \mathrm{d}\Omega^{(e)} \tag{4.2}$$

where constitutive matrix $\mathbf{D}^{(e)}\left(\mathbf{x}\right)$ is a function of the coordinates. In the original formulation the same shape functions are used for approximation of the displacement field and material inhomogeneity. However, from the numerical point of view nothing stands in the way of implementation of shape functions referring directly to the individual character of inhomogeneity, for instance power functions (Akai, Ootao and Tanigawa [8]) or exponential functions (Bagri, Eslami and Samsam-Shariat [14]). In the paper by Hernik and Ganczarski [102], in contrast with work by Woźniak et al. [190, 238] devoted to averaging techniques for a certain class of periodic functionally graded materials the inhomogeneity of which varies very slowly, it is shown that application of an averaging technique for zero-order approximation (Michalak *et al.* [158]) to the problem described by the Fourier heat conduction of highly oscillating coefficients reduces it to a homogeneous problem. Therefore a higher-order approximation based on the cosine variation of the thermal conductivity is proposed and numerically modelled by use of FEM code, having Kim and Paulino isoparametric graded finite elements built in. Comparison of results obtained by application of the present approach with classical FEM formulation and the Woźniak method for sampling a composite built of aluminium alloy Al and zirconia ZrO_2 partially stabilized by Y_2O_3, which exhibits drastically different magnitudes of thermal conductivity between adjacent components, is shown in Fig. 4.3

Wawrzynek and Ingraffea [234] developed the two-dimensional public domain FEM FRacture ANalysis Code FRANC–2D for implementing fracture

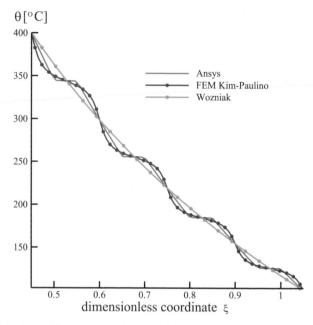

Fig. 4.3. Solution of heat conduction problem for periodic FGM (after Hernik and Ganczarski [102])

capabilities in both *isotropic* and *orthotropic* FGMs. The other possibility is to use *meshless methods* (Sladek, Sladek and Zhang [211]), rather than FE discretization.

Simple approximation of continuous unidirectional gradation in elasticity modulus, thermal expansion coefficient and thermal conductivity through exponential functions are proposed by Noda and Jin [166]

$$E = E_0 e^{\beta x} \qquad \alpha = \alpha_0 e^{\gamma x} \qquad \lambda = \lambda_0 e^{\delta x} \qquad (4.3)$$

In the case of ceramic–metal unidirectional composites a power function of the volume fraction of the metal phase is also frequently used (1.10–1.17), as shown in Fig. 1.4.

$$V_{\mathrm{m}} = \left(\frac{x}{l}\right)^m \qquad (4.4)$$

A power function of the volume fraction of the tungsten phase is used for the FGM–layer in the W–FGM–Cu composite by Ueda [219, 220, 220], Ueda and Gasik [223] and others.

When *orthotropic unidirectionally graded* material is considered, more general approximation formulae are used (Kim and Paulino [125]), e.g. for the components of the *plane elasticity tensor* it holds that:

$$E_{11}(x_1) = E_{11}^0 e^{\alpha x_1} \qquad E_{22}(x_1) = E_{22}^0 e^{\beta x_1} \qquad G_{12}(x_1) = G_{12}^0 e^{\gamma x_1} \qquad (4.5)$$

where E_{11}^0, E_{22}^0 and G_{12}^0 are elasticity moduli for a homogeneous (reference) material.

Three particular cases can also be considered:

(i) homogeneous orthotropic material: $(\alpha, \beta, \gamma) = (0, 0, 0)$,
(ii) orthotropic FGM with a proportional property variation:
$\quad (\alpha, \beta, \gamma) = (\alpha, \alpha, \alpha)$,
(iii) orthotropic FGM with a nonproportional property variation:
$\quad (\alpha \neq \beta \neq \gamma)$.

In case of thermal and mechanical application of FGMs, again three possibilities may occur:

(i) spatial gradation of mechanical properties only,
(ii) spatial gradation of thermal properties,
(iii) spatial gradation of both mechanical and thermal properties only.

When temperature-dependence of the properties is accounted for, in the case of a unidirectional temperature field $T(x_1)$ (4.5) should read

$$E_{11}(x_1, T) = E_{11}^0(T)e^{\alpha x_1} \quad E_{22}(x_1, T) = E_{22}^0(T)e^{\beta x_1}$$

$$G_{12}(x_1, T) = G_{12}^0(T)e^{\gamma x_1}$$

(4.6)

4.2 Application of the Concept of FGMs to Simple Thermo–elasto–plastic Structures

4.2.1 Study of CrN/W300 TBC System Under Cyclic Thermal Loading

Introduction

Motivation for the analysis of the TBC system, which consists of a thin *chromium nitride CrN layer* deposited on a tool steel substrate W300, is cyclic thermal loading locally applied at the piston engine ring surface. Thermal load is caused by frictional heating at the ring/cylinder contact zone induced by heat source due to local coating friction. A similar problem can be met when casting tools are applied, but in this case low-cycle thermal loads (of the other of some seconds) are induced, rather then high-cycle ones (of the order of microseconds) in the previous application.

Besides good wear resistance, mainly due to improved hardness, the chromium nitride CrN thin layer deposited on the exterior surface of the ring also serves as a *Thermal Barrier Coating TBC*, that should protect the steel W300 substrate from thermal ratchetting phenomena (Hones *et al.* [103]). Thermal ratchetting may result from the high-cycle heat fluxes entering the metallic substrate (the ring). The thin CrN layer is usually deposited at a high temperature condition during processing, which induces residual stresses in the layer, built-in during the cool down from the elevated fabrication temperature (Schulz et al. [202]). Depending on the deposition condition (temperature and pressure) and thickness of the CrN layer, usually ranging from 4 to 8 microns, the compressive residual stress in the coating is of magnitude between 1 GPa and 3 GPa, or sometimes even up to 4 GPa, as reported by Ebner [62]. In contrast to the tool steel W300, the properties of which are well reported in the literature, the properties of CrN are hardly recognized. They depend strongly on the deposition techniques, and are difficult to identify due to the small layer thickness, measured in microns. Additionally, as mentioned in Hones et al. [103], the properties of the chromium nitride layer should be considered as anisotropic and transversaly inhomogeneous, rather then isotropic and homogeneous. For the purpose of the following simulation we assume that the CrN layer is considered as pure elastic, isotropic, homogeneous ceramic material, whereas the temperature resistant tool steel W300 substrate is described by the elastic–plastic linear kinematic hardening rule (Oleksy and Skrzypek [171]). Study of the thermal response of the TBC system comprised of two approaches. The first is based on steady state stress–strain analysis, including the effect of residual stresses. The second consists in a full transient, time-dependent analysis to account for the plastic strains and thermal ratchetting phenomena in the steel substrate.

Composite properties

The ring is composed of the thin chromium nitride layer CrN deposited on the high temperature resistant tool steel W300. Mechanical and thermal properties of CrN are taken from Hones et al. [103], and Matweb [240]. Thermal dependence of the properties of this thin ceramic layer is not accounted for in the analysis (Table 4.1). Properties of the tool steel W300, mechanical and

Table 4.1. Material data for CrN

Material property	Magnitude
Elastic modulus	185 GPa
Poisson's ratio	0.3
Thermal expansion	0.0000007 K^{-1}
Thermal conductivity	12.0 $Wm^{-1}K^{-1}$
Specific heat	465 $Jkg^{-1}K^{-1}$
Mass density	7190 kgm^{-1}

thermal, are strongly temperature dependent as shown in Table 4.2, Table 4.3 and Table 4.4 (after Bohler [239]), such that temperature dependence has to be included in the analysis. The linear kinematic hardening model, based on the Huber–von Mises concept, is applied (Chaboche–Rousselier [38]).

$$F = \left(s_{ij} - \frac{2}{3} E_2 e_{ij}^{\mathrm{pl}} \right) \left(s_{ij} - \frac{2}{3} E_2 e_{ij}^{\mathrm{pl}} \right) - \frac{2}{3} \sigma_0^2 \qquad (4.7)$$

Table 4.2. Material data for W300

Temperature [°C]	E [GPa]	ν	ρ [kgm^{-3}]
20	224	0.3	7800
100	215	0.3	7800
200	206	0.3	7800
300	197	0.3	7800
400	188	0.3	7800
500	176	0.3	7800
600	165	0.3	7800
700	156	0.3	7800
800	145	0.3	7800

Table 4.3. Thermal properties for W300

Temperature [°C]	α [K^{-1} × 10^{-6}]	λ [Wm^{-1}K^{-1}]	c [Jkg^{-1}K^{-1}]
20	11.0	24.3	500
100	11.5	26.0	500
200	12.0	27.7	500
300	12.2	28.9	500
400	12.5	29.5	500
500	12.9	29.5	500
600	13.0	29.1	500
700	13.0	29.1	500
800	13.0	29.1	500

Boundary conditions

For the purpose of FE analysis by ANSYS, a 3 mm thick ring of radius 5 cm is modelled under plane stress conditions, and the rotational symmetry of the ring and axial symmetry of the 6 mm length heated zone are considered (Fig. 4.4a, b). For the thermal conditions, the heat pulse/cooling are uniformly applied over the 6 mm length heat affected zone, whereas the adiabatic condition $q_n = 0$ is used along the heat-free remaining portion of 3 μm thick CrN coating. When simulating low-cycle thermal fatigue testing in tool application a sawtooth temperature profile of constant amplitude and period $\Delta t = 2$ s

Table 4.4. W300 properties for linear hardening (4.7)

Temperature	σ_0 [MPa]	E_2 [MPa]
20	1000	800
200	930	800
300	900	800
400	800	727
500	650	615
600	450	375
700	250	215
800	50	175

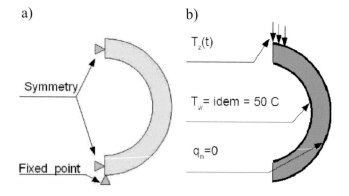

Fig. 4.4. Boundary conditions (a), loading (b)

is used. For high-cycle thermal ratchetting simulation in a piston ring application the corresponding $t_{\mathrm{pulse}}/t_{\mathrm{brake}} = 4$ ms/16 ms is assumed. Two types
of thermal conditions are investigated at the interior (cool) edge of the ring.
In the case of steady-state analysis, a constant temperature T_1 is held at the
cool edge. In the case of transient (high-cycle) analysis the heat flux entering
the heat affected zone of the CrN layer propagates inwards in the substrate
and, for the metal, is asummed to have the same intensity as along the cool
interior surface of the ring (see Fig. 4.4). Only the heat flux due to thermal
conductivity is accounted for; neither convection nor radiation are included.
Hence, the *heat balance equation* (4.7), with internal heat sources neglected,
is

$$\frac{\partial \left(\rho \left(\mathbf{x}, t \right) c \left(\mathbf{x}, T \right) \right)}{\partial t} + \operatorname{div} \mathbf{q} \left(\mathbf{x}, t \right) - Q \left(\mathbf{x}, t \right) = 0 \qquad (4.8)$$

where the heat flux is given by the Fourier law

$$\mathbf{q} \left(\mathbf{x}, t \right) = -\lambda \left(\mathbf{x}, T \right) \operatorname{\mathbf{grad}} T \left(\mathbf{x}, t \right) \qquad (4.9)$$

Numerical simulation

Simulation of the fabrication phase to yield the *residual stress* in the CrN/W300 composite structure is done in two ways. First, when the fabrication phase is thermally simulated, a uniform stress-free heat-up of the composite structure, starting from the initial temperature $T_1 = 50°C$ to the fabrication temperature T_2, ranging from $500°C$ to $800°C$, is modelled by the use of "killed" elements in the FE mesh. Then, when the cool-down phase begins, true ("alive") elements are used, so that residual stresses occur: compressive, ranging from 0.8 GPa to 1.5 GPa in the CrN coating (ceramic) and insignificant tensile stresses in the W300 substrate. In the case of a steady-state analysis the cool-down phase is performed as long as the ultimate temperature at the heat affected zone T_2 drops to the initial (reference) temperature T_1.

Second, simplified simulation of residual stresses consists in providing a fixed mean residual stress in the CrN coating, of a value ranging from 2 GPa to 4 GPa, whereas the corresponding residual stresses in the substrate are calculated from the equilibrium. When the transient analysis of short heat pulses of constant magnitude is performed, after the pulse/break cycle is finished the ultimate temperature does not reach the initial temperature, such that an incomplete cool-down of the composite structure leads to thermal ratchetting phenomena on the consecutive thermal cycles.

Steady-state analysis under fabrication/working thermal cycling 50°C–500°C–50°C

Steady-state simulation is performed using sawtooth temperature control. The fabrication cycle $50°C–500°C–50°C$ comprises the stress-free uniform heat-up of the structure and stress-active cool-down. Consecutive heat pulse/break cycling are also temperature controlled ranging between $500°C$ (hot) and $50°C$ (cool). Steady-state process control by sawtooth temperature cycles is realistic as long as temperature fluctuations are slow enough to ensure complete cool-down of the structure to unchanged ultimate temperature $T_1 = const$, at each thermal cycle. In Fig. 4.5 a sequence of the circumferential stress across the depth, with time parameter during the steady-state heat-up/cool-down $50°C–500°C–50°C$, is presented. The highest compressive stress on the CrN surface reaches -0.8 GPa at the end of a cool-down fabrication phase. When temperature at the heat affected zone increases, the compressive stress in CrN decreases, whereas the tensile stress in W300 substrate increases for the sake of equilibrium. However, stresses in the metal are not high enough to initiate yielding, such that the whole process is reversible elastic on the next loading cycle, as shown at a point in the metal neighbouring the ceramic layer in Fig. 4.6.

Note the nonlinear stress–stress path, which results from cyclically changing temperature-dependent properties of the metal. Evolution of the stress σ_φ with time during the fabrication phase and first working thermal cycle is

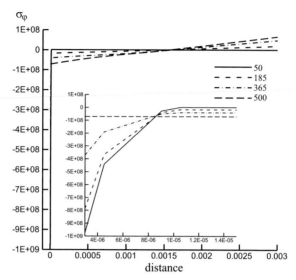

Fig. 4.5. Hoop stress as a function of distance

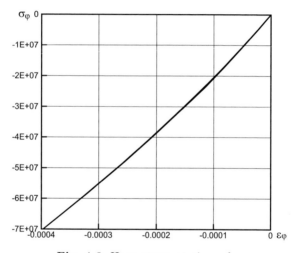

Fig. 4.6. Hoop stress–strain cycles

shown in Fig. 5. Period 1 corresponds to stress-free heat-up, whereas period 2 represents the residual stress in metal with time arising from stress-active cool-down of the structure from fabrication temperature. Periods 3 and 4 correspond to heat-up (50°C–500°C) and cool-down (500°C–50°C) on first thermal cycling. The highest compressive stress in metal reaches -72 MPa, which results from the thermal expansion coefficients mismatch between CrN $(0.7 \times 10^{-6} \mathrm{K}^{-1})$ and W300 $(11.25 \times 10^{-6} \mathrm{K}^{-1})$ at temperature 50°C (Fig. 4.7).

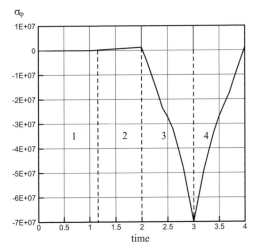

Fig. 4.7. Evolution of hoop stress

Steady-state analysis of fixed residual stress 4 GPa under thermal cycling 50°C–800°C–50°C

Fixed mean value 4 GPa of compressive residual stress in CrN was assumed for *cyclic thermal loading* controlled by reverse temperatures 50°C–800°C–50°C, etc. In the case considered a severe stress gradient occurs in the metallic zone neighbouring the CrN coating as shown in Fig. 4.8. During the heat-up

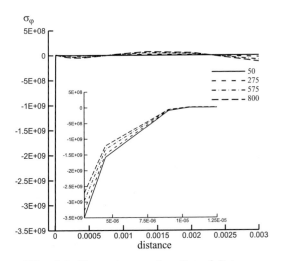

Fig. 4.8. Hoop stress as function of distance

working phase up to 800°C, plasticization in metal is observed at compressive stress -80 MPa, to finally reach the value -60 MPa by the end of heating (points 2'and 2 in Fig. 4.9). Consecutive changes of stress on the next thermal

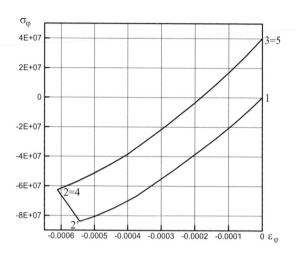

Fig. 4.9. Hoop stress–strain cycles

cycle are purely elastic (points 3–4–5, etc.). The corresponding diagram of stress in a plasticized metal zone with time is presented in Fig. 4.10, where elastic–plastic heat-up phase (1–2'–2) is followed by tensile stress increase in metal (2–3) on cool-down, and pure elastic consecutive thermal cycling (3–4–5; etc.). Due to steady-state conditions used for analysis, a complete cool-down (800°C–50°C) occurs at each consecutive loading cycle, such that thermal ratchetting is not observed.

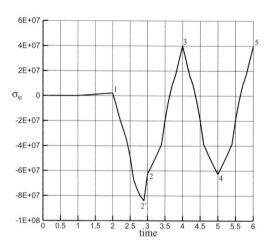

Fig. 4.10. Evolution of hoop stress

Ratchetting analysis under high-cycle heat pulse/break

Transient thermal cycling analysis is performed in this section assuming $t_{\text{pulse}}/t_{\text{break}} = 4\ \text{ms}/16\ \text{ms}$, which makes analysis closer to the piston-ring car engine working condition. Profiles of high-cyclic heat fluxes are shown in Fig. 4.11. The problem is controlled now by a constant heat pulse ampli-

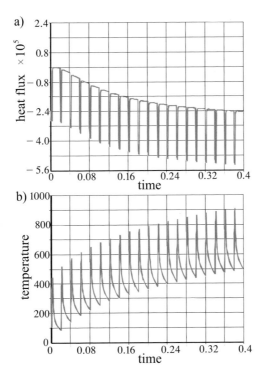

Fig. 4.11. Excitation by heat flux and temperature response

tude over consecutive thermal cycles. For the initial condition a fixed residual stress in the CrN layer of -4 GPa is assumed. During the first working heat cycle, heat flux intensity is chosen in such a way that it cause temperature change from $100°C$ to $500°C$. The full heat balance equation (4.8) is solved at each thermal cycle, however, during the cool-down after the heat pulse is removed, temperature drop is retarded. As consequence, when the frictionless break is finished, a new heat pulse enters the CrN/W300 system at increased initial temperature for the next thermal cycle. Obviously, when heat pulse intensity remains unchanged over the whole thermal cycling programme, resulting reverse point temperature T_2 also increases cycle-by-cycle, such that *thermal ratchetting* is observed. After a number of cycles, the temperature in steel at reverse points approaches the saturation level, and the accompanying stress–strain diagram is shown in Fig. 4.12.

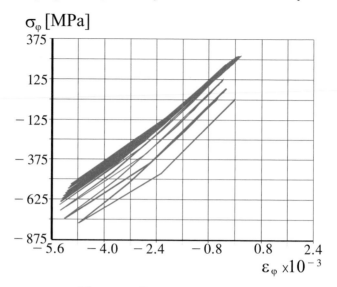

Fig. 4.12. Stress–strain diagram

4.2.2 Stability Analysis Against the Hot-spots in a FG Thermo–elastic Brake Disc

Introduction

An increase in the speed of vehicles causes an increase of the dissipated energy in braking systems. One consequence of this is the frequent occurrence of *hot-spots*. Experimental investigations using an infrared camera allow one to trace this phenomenon not only on a TGV brake disc but also on a car clutch disc (see Fig. 4.13). Explanation of the mechanisms of the occurrence

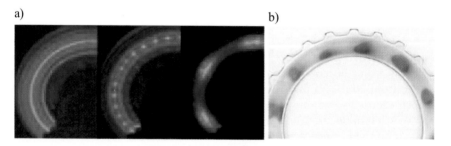

Fig. 4.13. Hot-spots registered on: (a) TGV brake disc [116], (b) clutch disc [242]

of hot-spots is based on the thermomechanical analysis, taking into account contact between disc and pads. During intensive braking an amount of heat generated on the friction surfaces causes the occurrence of a global buckling

mode of the disc and consequent reduction of the contact surfaces with very
high local temperatures. Explanation of this mechanism is still under discussion; several papers have already been published by Fan *et al.* [71], Du *et al.* [59] and Dufrénoy and Weichert [61]. In what follows, the results of the *hot-spots* analysis by the concept of FGM, is proposed after Hernik [101].

Formulation of the problem

The aim of the present chapter is the numerical modelling of a FGM brake disc
to prevent the occurrence of hot-spots. Macroscopic (effective) thermomechanical properties of a two-phase (metal–ceramic) composite (MMC) depends on
the volume fraction of both components. The volume fraction is subjected to
continuous change along certain coordinate and causes a smooth change of
effective properties of the MMC, as shown in Fig. 4.14. The volume fraction

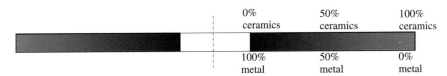

| 0% ceramics | 50% ceramics | 100% ceramics |
| 100% metal | 50% metal | 0% metal |

Fig. 4.14. MMC brake disk after Hernik [101]

is described by the *mixture rules* given by (1.8) and (1.9), hence in accordance with them the following property gradations for thermal conductivity
coefficient and modulus of elasticity are proposed:

- the first based on exponential functions (4.3) after Noda and Jin [166]

$$\lambda\left(r\right) = \lambda_0 \exp\left(k_1 \frac{r}{r_2}\right) \qquad E\left(r\right) = E_0 \exp\left(k_2 \frac{r}{r_2}\right) \tag{4.10}$$

- the other based on a power function (4.4) proposed by Eslami *et al.* [69, 70]
  and Ootao *et al.* [174]

$$\lambda\left(r\right) = \lambda_0 \left(\frac{r}{r_2}\right)^{m_1} \qquad E\left(r\right) = E_0 \left(\frac{r}{r_2}\right)^{m_2} \tag{4.11}$$

The model of the brake disc is the Kirchhoff–Love plate (see Fig. 4.15). It is
assumed that all thermal and mechanical loadings satisfy simplified conditions
of full rotational symmetry. Centrifugal loadings in both radial and hoop
direction fulfil this assumption exactly, however, in the case of friction forces
generating a temperature field, an appropriate homogenization to the whole
circumference is necessary. The braking is treated as a process occurring under
constant negative acceleration $\varepsilon =$const.

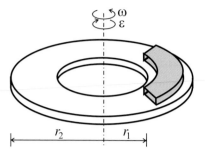

Fig. 4.15. Model of brake disc

The mathematical model of the brake disc is established by the following four equations:

1. The *heat transfer equation* of the Fourier type (2.1)

$$\frac{1}{r}\frac{d}{dr}\left(r\lambda\left(r\right)\frac{dT}{dr}\right) + \dot{q}_v = \varrho c\dot{T} \tag{4.12}$$

in which the intensity of inner heat sources corresponds either to heat generation in the friction zone or to cooling-down beyond the contact zone

$$\dot{q}_v = \begin{cases} 2\dfrac{\mu p\omega\eta}{h}r & \text{in friction zone} \\ -2\dfrac{\beta(\omega)}{h}\left(T - T_\infty\right) & \text{beyond friction zone} \end{cases} \tag{4.13}$$

where h is the disc thickness, p is pressure, μ denotes friction coefficient, β is the coefficient of free convection, η is the coefficient of efficiency and T_∞ stands for temperature of cooling air outside the boundary layer.

2. The first equation of *membrane state for radial displacement*

$$\mathcal{B}\overline{\nabla}_1^2 u + \frac{d\mathcal{B}}{dr}\left(\frac{du}{dr} + \nu\frac{u}{r}\right) + \varrho\omega^2 h = (1+\nu)\,\alpha\left[\mathcal{B}\frac{dT}{dr} + \frac{d\mathcal{B}}{dr}\left(T - T_{\text{ref}}\right)\right] \tag{4.14}$$

where T_{ref} is a reference temperature.

3. The second equation of *membrane state for hoop displacement*

$$\begin{aligned} &\frac{1-\nu}{2}\mathcal{B}\overline{\nabla}_1^2 v + \frac{1-\nu}{2}\frac{d\mathcal{B}}{dr}\left(\frac{dv}{dr} - \frac{v}{r}\right) + \varrho\varepsilon h = 0 \quad \text{beyond friction zone} \\ &\frac{1-\nu}{2}\mathcal{B}\frac{d^2 v}{dr^2} + \left(\frac{1}{r} + \frac{d\mathcal{B}}{dr}\right)N_0 + \varrho\varepsilon h = 0 \qquad \text{in friction zone} \end{aligned} \tag{4.15}$$

where N_0 denotes the friction force calculated as integral

$$N_0 = \int_{r_3}^{r_4} 2p\mu\,dr \tag{4.16}$$

taken over the width of the brake pad, whereas the differential operator is given by the following formula:

$$\overline{\nabla}_1^2(\ldots) = \frac{\mathrm{d}^2 \ldots}{\mathrm{d}r^2} + \frac{\mathrm{d} \ldots}{\mathrm{d}r} - \frac{\ldots}{r^2} \tag{4.17}$$

4. System of two equations of *bending state* proposed by Fedehofer and Egger [73] performing axial displacement as combination $w(r,\varphi) = f_1(r)\sin(k\varphi) + f_2(r)\cos(k\varphi)$

$$\overline{\nabla}_k^2\left(\mathcal{D}\overline{\nabla}_k^2 f_{1/2}\right) - \frac{1}{\Lambda}\left[N_r\frac{\mathrm{d}^2 f_{1/2}}{\mathrm{d}r^2} + \frac{N_\varphi}{r}\left(\frac{\mathrm{d}f_{1/2}}{\mathrm{d}r} - \frac{k^2}{r}f_{1/2}\right)\right.$$
$$\mp \frac{2kN_{r\varphi}}{r}\left(\frac{\mathrm{d}f_{2/1}}{\mathrm{d}r} - \frac{f_{2/1}}{r}\right) - \varrho\omega^2 h\frac{\mathrm{d}f_{1/2}}{\mathrm{d}r} \pm \varrho\varepsilon h\frac{k}{r}f_{2/1}\bigg] - q_z = 0 \tag{4.18}$$

in which k stands for subsequent buckling mode, whereas the differential operator is an extension of one given by (4.17)

$$\overline{\nabla}_k^2(\ldots) = \frac{\mathrm{d}^2 \ldots}{\mathrm{d}r^2} + \frac{\mathrm{d} \ldots}{\mathrm{d}r} - k^2\frac{\ldots}{r^2} \tag{4.19}$$

and definitions of *membrane and bending stiffnesses* include change of elastic modulus along r, whereas ν is considered constant

$$\mathcal{B} = \frac{E(r)h}{1-\nu^2}, \qquad \mathcal{D} = \frac{E(r)h^3}{12(1-\nu^2)} \tag{4.20}$$

During intensive braking accompanied by high temperature, when a buckling mode appears, a pair of brake pads clamp the brake disc, stabilizing it against buckling. This effect is taken into consideration in (4.18) by the additional term q_z modelling the elastic contact with the Winkler foundation

$$q_z = \begin{cases} 0 & \text{beyond friction zone} \\ -\gamma f_{1,2} & \text{in friction zone} \end{cases} \tag{4.21}$$

Coefficient of Winkler foundation is calculated from the empirical formula

$$\gamma = 2.7 \times 10^4 \frac{\Delta V_N}{l_p^2} \tag{4.22}$$

combining its magnitude with the volume change of brake fluid ΔV_N and the length of flexible cords l_p.

The system of differential equations ((4.12), (4.14), (4.15) and (4.18)) describes the problem of stability, where $1/\Lambda$ stands for the Lagrange multiplier referring to a set of loadings. In the case of analysis of structure subjected to a set of n independent and conservative loadings, the existence of a critical state is determined by the approximate *Schaefer–Papkovich condition*

$$\sum_{i=1}^n \frac{P_i}{P_{i_{cr}}} = \Lambda = 1 \tag{4.23}$$

It is worth noting that the derived system of differential equations ((4.12), (4.14), (4.15) and (4.18)) is quasi-coupled (temperature occurs as the right-hand side in the equations of mechanical state but there are no generalized forces in the heat transfer equation), hence the application of the *stagger algorithm* is favourable.

Boundary conditions

Uniqueness of the solution of the problem described by the system of differential equations ((4.12), (4.14), (4.15) and (4.18)) requires formulation of an appropriate set of boundary conditions.

In the case of the Fourier equation (4.12), the following set of initial and boundary conditions is considered

$$T(r,0) = T_{\text{ref}} \qquad -\lambda \frac{\mathrm{d}T(r,\tau)}{\mathrm{d}r}\bigg|_{r_1} = -q_0 \qquad -\lambda \frac{\mathrm{d}T(r,\tau)}{\mathrm{d}r}\bigg|_{r_2} = 0 \qquad (4.24)$$

where q_0 stands for heat flux transferred to the hub. It is assumed that the brake disc is fixed at the inner edge to the hub, whereas its outer edge is free so the boundary conditions for the equations of mechanical state ((4.14), (4.15) and (4.18)) are as follows

$$
\begin{aligned}
u\big|_{r_1} &= 0 & N_r &= B\left[\frac{\mathrm{d}u}{\mathrm{d}r} + \nu\frac{u}{r} - (1+\nu)\alpha(T - T_{\text{ref}})\right]\bigg|_{r_2} = 0 \\
v\big|_{r_1} &= 0 & N_{r\varphi} &= \frac{1-\nu}{2}B\left(\frac{\mathrm{d}v}{\mathrm{d}r} + \frac{v}{r}\right)\bigg|_{r_2} = 0 \\
f_i\big|_{r_1} &= 0 & M_r &= \left[\frac{\mathrm{d}^2 f_i}{\mathrm{d}r^2} + \frac{\nu}{r}\left(\frac{\mathrm{d}f_i}{\mathrm{d}r} - k^2\frac{f_i}{r}\right)\right]\bigg|_{r_2} = 0 \\
\frac{\mathrm{d}f_i}{\mathrm{d}r}\bigg|_{r_1} &= 0 & \overline{Q}_r &= \left[\frac{\mathrm{d}^3 f_i}{\mathrm{d}r^3} + \frac{1}{r}\frac{\mathrm{d}^2 f_i}{\mathrm{d}r^2} - \frac{1 + k^2(2-\nu)}{r^2}\frac{\mathrm{d}f_i}{\mathrm{d}r} + \right. \\
& & &\left. \frac{k^2(3-\nu)}{r^3}f_i\right]\bigg|_{r_2} = 0
\end{aligned}
\qquad (4.25)
$$

Numerical algorithm

Starting the numerical analysis of the system of differential equations ((4.12), (4.14), (4.15) and (4.18)) note that the first of them is a partial differential equation of parabolic type, which requires a separate treatment to the remaining ordinary differential equations. From the numerical point of view initial-boundary value problem (4.12) and (4.24) can be classified as a diffusive flux-conservative initial value problem with prototypical equation in the form

$$\frac{\partial T}{\partial \tau} = \frac{\partial}{\partial r}\left[F(r)\frac{\partial T}{\partial r}\right] \qquad (4.26)$$

where $F(r)$ is the diffusion coefficient. Considering first a simplified case when F is constant, when the fully implicit or backward time differencing scheme is applied, we get the tridiagonal system

$$aT_{j-1}^k + (1 - 2a)T_j^k + aT_{j+1}^k = T_j^{k+1} \qquad (4.27)$$

where term $a = F\Delta\tau/(\Delta r)^2$ is the Fourier criterion number, the magnitude of which determines whether the scheme is unconditionally stable or not. The von Neumann stability analysis requires $a \leq 0.5$ or, equivalently, limitation of time step $\Delta\tau \leq 0.5(\Delta r)^2/a$.

In order to derive an appropriate stability condition for the original equation (4.12) the fully implicit differencing scheme is applied first to this version of the equation, which is valid for the friction zone,

$$
\frac{\Delta\tau}{c\varrho}\left[\frac{1}{2\Delta r}\left(\frac{\lambda}{r_j} + \frac{d\lambda}{dr}\right) - \frac{\lambda}{(\Delta r)^2}\right]T_{j-1}^{k+1} + \left(\frac{2\lambda\Delta\tau}{(\Delta r)^2 c\varrho} + 1\right)T_j^{k+1} +
$$
$$
- \frac{\Delta\tau}{c\varrho}\left[\frac{1}{2\Delta r}\left(\frac{\lambda}{r_j} + \frac{d\lambda}{dr}\right) + \frac{\lambda}{(\Delta r)^2}\right]T_{j+1}^{k+1} - 2\frac{\mu p\omega\eta}{h}r_j = T_j^k \quad (4.28)
$$

leading to the stability condition (Kapturkiewicz [123])

$$
\Delta\tau \leq \frac{\varrho c(\Delta r)^2}{2\lambda} \quad (4.29)
$$

Nevertheless, the above estimation turns out to be insufficient because $\lambda = \lambda(r)$, hence the extended stability condition is used (Press et al. [183])

$$
\Delta\tau \leq \min_j\left[\frac{\varrho c\,(\Delta r)^2}{2\lambda\,(r_j)}\right] \quad (4.30)
$$

If the fully implicit differencing scheme is applied to the version of Eq. (4.12), which is valid beyond the friction zone, we get

$$
\frac{\Delta\tau}{c\varrho}\left[\frac{1}{2\Delta r}\left(\frac{\lambda}{r_j} + \frac{d\lambda}{dr}\right) - \frac{\lambda}{(\Delta r)^2}\right]T_{j-1}^{k+1} +
$$
$$
\left[\left(\frac{2\lambda}{(\Delta r)^2} + 2\frac{\beta(\omega)}{h}\right)\frac{\Delta\tau}{c\varrho} + 1\right]T_j^{k+1} -
$$
$$
\frac{\Delta\tau}{c_v\varrho}\left[\frac{1}{2\Delta r}\left(\frac{\lambda}{r_j} + \frac{d\lambda}{dr}\right) + \frac{\lambda}{(\Delta r)^2}\right]T_{j+1}^{k+1} - 2\frac{\beta(\omega)}{h}T_\infty = T_j^k \quad (4.31)
$$

giving the analogous stability condition

$$
\Delta\tau \leq \min_j\left[\frac{c\varrho}{\frac{2\lambda(r_j)}{(\Delta r)^2} + 2\frac{\beta(\omega)}{h}}\right] \quad (4.32)
$$

which reduces to (4.30) for $\beta(\omega) = 0$.

Numerical integration of the problem takes advantage of a step-by-step procedure and the shooting method (Press et al. [183]). First, the problem given by the system of ordinary differential equations ((4.12), (4.14), (4.15) and (4.18)) is performed in standard form for two point boundary value problems by inserting new variables

$$x = r \qquad y_1 = T \qquad y_2 = \frac{\mathrm{d}T}{\mathrm{d}r} \qquad y_3 = u \qquad y_4 = \frac{\mathrm{d}u}{\mathrm{d}r}$$

$$y_5 = v \qquad y_6 = \frac{\mathrm{d}u}{\mathrm{d}r} \qquad y_7 = f_1 \qquad y_8 = \frac{\mathrm{d}f_1}{\mathrm{d}r} \qquad y_9 = \frac{\mathrm{d}^2 f_1}{\mathrm{d}r^2} \tag{4.33}$$

$$y_{10} = \frac{\mathrm{d}^3 f_1}{\mathrm{d}r^3} \quad y_{11} = f_2 \quad y_{12} = \frac{\mathrm{d}f_2}{\mathrm{d}r} \quad y_{13} = \frac{\mathrm{d}^2 f_2}{\mathrm{d}r^2} \quad y_{14} = \frac{\mathrm{d}^3 f_2}{\mathrm{d}r^3}$$

which means that it is written as coupled sets of $N = 14$ first-order equations

$$\frac{\mathrm{d}y_i\,(x)}{\mathrm{d}x} = g_i\,(x, y_1, \dots, y_N, \Lambda) \tag{4.34}$$

in which the Lagrange multiplier appears as a new 15th dependent variable

$$y_{15} \equiv \Lambda \tag{4.35}$$

satisfying the additional differential equation

$$\frac{\mathrm{d}y_{15}}{\mathrm{d}x} = 0 \tag{4.36}$$

At starting point x_1 the solution is supposed to satisfy n_1 boundary conditions

$$B_{1j}\,(x_1, y_1, \dots, y_N) = 0 \qquad j = 1, \dots, n_1 \tag{4.37}$$

while at the end point x_2 it is supposed to satisfy the remaining $n_2 = N - n_1 = 7$ boundary conditions

$$B_{2k}\,(x_2, y_1, \dots, y_N) = 0 \qquad k = 1, \dots, n_2 \tag{4.38}$$

Numerical implementation of the shooting method is based on a multidimensional, globally convergent *Newton–Raphson method* seeking solution of n_2 functions including n_2 variables (Press *et al.* [183]). At the start point x_1 there are N starting values y_i, but only n_1 boundary conditions, hence there are $n_2 = N - n_1$ freely specifiable starting values. Assuming that these freely specifiable starting values are the components of a vector $\mathbf{V}$ belonging to n_2 dimensional linear space, the use of subroutine load.for generates a complete set of N starting values $\mathbf{y}$, satisfying the boundary conditions at x_1 supplemented by vector $\mathbf{V}$

$$y_i\,(x, V_1, \dots, V_{n_2}) = 0 \quad i = 1, 2, \dots, N \tag{4.39}$$

This means that choice of a certain vector $\mathbf{V}$ defines the starting vector $\mathbf{y}\,(x_1)$, which by integration of the system (4.34) leads to a solution $\mathbf{y}\,(x_2)$. If it is defined, the discrepancy vector $\mathbf{F}$ at x_2 belonging to n_2 dimensional linear space, whose components measure how far the solution is from the boundary conditions at x_2

$$F_k = B_{2k} (x_2, \mathbf{y}) \quad k = 1, 2, \ldots, n_2, \tag{4.40}$$

The use of subroutine `score.for` changes the N-dimensional vector of ending values $\mathbf{y}(x_2)$ into an n_2-dimensional vector of discrepancies $\mathbf{F}$. In this way, from the Newton–Raphson method point of view, the problem is reduced to searching for a vector value of $\mathbf{V}$ that zeros the vector value of $\mathbf{F}$. The globally convergent Newton's method implemented in routine `newt` solves the set of n_2 linear equations

$$\mathbf{J} \cdot \delta \mathbf{V} = -\mathbf{F} \tag{4.41}$$

and next adds the correction

$$\mathbf{V}^{\text{new}} = \mathbf{V}^{\text{old}} + \delta \mathbf{V} \tag{4.42}$$

The Jacobian matrix $\mathbf{J}$ has the following form

$$J_{ij} = \frac{\partial F_i}{\partial V_j} \tag{4.43}$$

which may be useless to compute all partial derivatives analytically in a general case. Therefore, evaluation of the Jacobian is done automatically in the routine `fdjac.for`, calculating it approximately

$$\frac{\partial F_i}{\partial V_j} \simeq \frac{F_i (V_1, V_2, \ldots, V_j + \Delta V_j, \ldots)}{\Delta V_j} \tag{4.44}$$

Results

All numerical examples deal with brake discs made of stainless steel ASTM-321 (rolled 18 Cr, 8 Ni, 0.45 Si, 0.4 Mn, 0.1 C, Ti/Nb stabilized, austenitic, annealed at 1070°C, air cooled) having material properties and other starting parameters as Odqvist [175], shown in Table 4.5. Geometric parameters of

Table 4.5. Material properties and other starting parameters of brake disc made of stainless steel ASTM-321; after Odqvist [175]

E_0 [GPa]	σ_0 [MPa]	ν [–]	ϱ [kg/m^3]	α [1/K]	λ_0 [W/mK]
170	120	0.3	7850	1.85×10^{-5}	20
β [W/m^{2}K]	q_0 [W/m^2]	η [–]	T_∞ [°C]	T_{ref} [°C]	T_0 [°C]
7÷50	30	10^{-3}	20	20	2×10^3
c [J/kgK]	ε [1/s^2]	μ [–]	ω_0 [1/s]	p [MPa]	γ [MPa/m]
478	7	0.4	81.71	0.4	870

brake disc installed in a sport car are as detailed in Dufrénoy and Weichert [61], Panier *at al.* [177] are shown in Fig. 4.16 and Table 4.6.

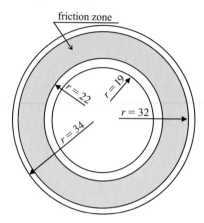

Fig. 4.16. Brake disc geometry; after Dufrénoy and Weichert [61], Panier *at al.* [177]

Table 4.6. Geometric parameters of brake disc; after Dufrénoy and Weichert [61], Panier *at al.* [177]

r_1 [m]	r_2 [m]	r_3 [m]	r_4 [m]	h [m]
0.19	0.34	0.22	0.32	0.045

Instability of a homogeneous brake disc

A solution to the reference problem of a homogeneous brake disc (E =const, λ =const) made of the stainless steel ASTM-321 follows experimental results by Dufrénoy and Weichert [61] and Panier *at al.* [177]. The temperature distribution shown in Fig. 4.17 begins with temperature T_1 for the time instant $\tau = \tau_0$, and temperatures T_2 through T_7 refer to subsequent time steps for time interval $d\tau = 0.15$ s. Detailed analysis reveals clearly the central zone of heat generation and two zones of cooling down located near inner and outer boundaries. The amount of heat generated decreases with the angular velocity ω, as a consequence of change in character of convection from a forced to a free process. Distribution of the generalized hoop stress N_φ, which is directly responsible for global loss of stability of the brake disc, since it turns out to be a single component of compressive nature, is shown in Fig. 4.18. Two sampling buckling modes ($k = 3$ and $k = 6$) are presented in Fig. 4.19. The first eight eigenvalues of the homogeneous brake disc are in the range $\Lambda_{1,...,9} \in [0.0378 \div 0.1169]$ and cause buckling modes since, according to Schaefer–Papkovich condition (4.23), they are smaller than unity.

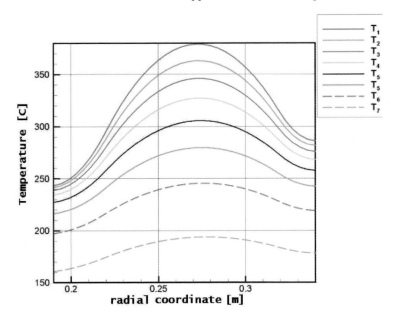

Fig. 4.17. Temperature distribution in homogeneous brake disc

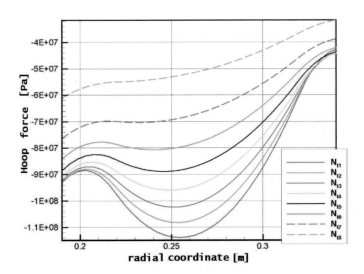

Fig. 4.18. Generalized hoop stress distribution in homogeneous brake disc

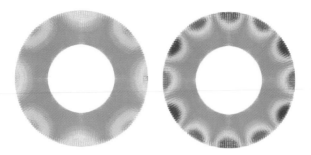

Fig. 4.19. Buckling modes of homogeneous brake disc for $k = 3$ and $k = 6$

Stability of a FG brake disc

Application of functional gradation using ceramic, according to rules (4.10) and (4.11), leads to changes in the distribution of λ and E, as shown in Fig. 4.20. Maximum temperature of the FGM brake disc is essentially lower in

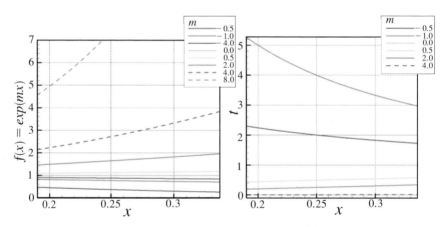

Fig. 4.20. Approximation functions of FGM brake disc: (a) exponential functions, (b) power functions; after Hernik [101]

comparison to the reference solution (cf. Fig. 4.21), which is a consequence of more intensive cooling down caused by a decrease of the thermal conductivity coefficient. In the case of distribution of the generalized hoop stress, the influence of functional gradation by the ceramic is also advantageous, since it reduces stress magnitude and, simultaneously, essentially improves stability of the brake disc (cf. Fig. 4.22). In the example under consideration the first eight eigenvalues are shifted to the safe range $\Lambda_{1,\dots,9} \in [35.426 \div 128, 894]$ defined by Schaefer–Papkovich condition (4.23).

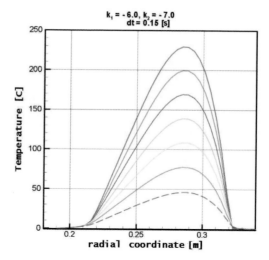

Fig. 4.21. Temperature distribution in FGM brake disc

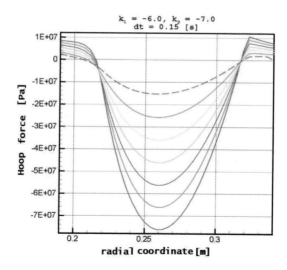

Fig. 4.22. Hoop stress distribution in FG brake disc

4.2.3 Modelling of FGM Al/ZrO$_2$+Y$_2$O$_3$ Engine Pistons

Introduction

The piston is one of the main parts of an engine. Its basic role is to transfer force from the combustion gases to a connecting rod. Apart from the gas pressure the piston is subjected to high temperature. Existence of these two loadings requires application of light alloys, mainly based on Al, Mg or Ti, in the production of pistons, since they combine low weight with high toughness and advantageous thermal conductivity. The real challenge for engineers are pistons made of FGM metal–ceramic composites (Fig. 4.23).

Fig. 4.23. Modern pistons made of FGM metal–ceramic composites

Experiments on the lifetime of pistons with and without a ceramic coating (TBC) done by HiPeCo confirm the distinct superiority of the former. In a series of experiments the insulation properties of pistons were examined by continuous heating and simultaneous measurement of temperature (Fig. 4.24). Temperature was measured for 10 min at 30 s intervals. After 5.5 min the

Fig. 4.24. Experiment heating the bottom of a piston, performed by HiPeCo

piston bottom without a ceramic coating was burnt out, whereas the piston bottom with ceramic coating revealed only a slight change of colour after 10 min (Fig. 4.25).

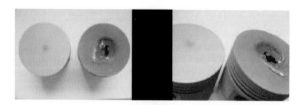

Fig. 4.25. Complete burning out of piston bottom without ceramic coating and slight change of colour of piston bottom with ceramic coating; after HiPeCo

In order to examine failure between coating and substrate the piston bottom was subjected to a further increase of temperature and simultaneous decrease of distance to the front of the flame. The surface of the ceramic coating did not exhibit any cracks or bulges even in places where the substrate material was completely melted (Table 4.7).

Table 4.7. Results of lifetime experiments on pistons with and without ceramic coating (after HiPeCo)

Time [min]	Temperature [°C] no coating	coating	Time [min]	Temperature [°C] no coating	coating	Time [min]	Temperature [°C] no coating	coating
0	18	18	3.5	415	315	7		425
0.5	115	95	4	440	340	7.5		435
1	185	145	4.5	455	365	8		450
1.5	240	191	5	463	380	8.5		455
2	305	230	5.5	470	390	9		460
2.5	345	270	6		400	9.5		465
3	385	300	6.5		408	10		468

Assumptions

For the numerical modelling of FGM engine pistons we follow the method presented by Cegielski [32].

The piston bottom made of FGM is treated as a thick plate satisfying the general equations of thermoelasticity (Fig. 4.26). The general assumptions of

Fig. 4.26. Piston made of FGM

rotational symmetry of both temperature and displacement fields hold

$$\partial T/\partial\varphi = 0, \qquad \partial \mathbf{u}/\partial\varphi = 0 \qquad (4.45)$$

where T denotes temperature, φ is hoop coordinate and $\mathbf{u}$ denotes displacement vector. Additionally, the linearized total strain tensor is decomposed into the elastic and thermal parts

$$\varepsilon = \varepsilon^{\mathrm{e}} + \varepsilon^{\mathrm{th}}, \qquad \varepsilon^{\mathrm{th}} = \mathbf{1}\alpha\theta, \qquad \theta = T - T_{\mathrm{ref}} \qquad (4.46)$$

where $\theta = T - T_{\mathrm{ref}}$ denotes the difference between actual and reference temperatures.

Heat transport in FGM plate

A general description of heat transfer in an isotropic but inhomogeneous FGM requires application of the Tanigawa approach [217] in which the coefficient of thermal conductivity λ depends on particle location and it appears under the divergence operator in the stationary Fourier's equation

$$\mathrm{div}\,[\lambda(z)\,\mathbf{grad}T(r,z)] = 0 \qquad (4.47)$$

For the piston bottom in a combustion engine the natural material gradation coincides with the transverse direction (Fig. 4.27), therefore (4.47), rewritten in the cylindrical coordinate system, takes the following form:

$$\frac{\partial^2 T}{\partial r^2} + \frac{1}{r}\frac{\partial T}{\partial r} + \frac{1}{\lambda}\frac{\partial \lambda}{\partial z}\frac{\partial T}{\partial z} + \frac{\partial^2 T}{\partial z^2} = 0 \qquad (4.48)$$

The magnitude of the thermal conductivity λ depends on the local volume

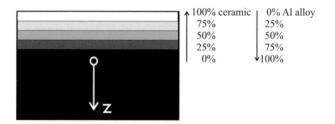

Fig. 4.27. Piston wall made of Al alloy functionally graded by ceramic

fraction of both FGM constitutents and is described by Noda's approximation [166] (4.3), which, substituted into (4.48), leads to the final form of *Fourier's equation*

$$\frac{\partial^2 T}{\partial r^2} + \frac{1}{r}\frac{\partial T}{\partial r} + k_1\frac{\partial T}{\partial z} + \frac{\partial^2 T}{\partial z^2} = 0 \qquad (4.49)$$

Uniqueness of solution of the partial differential equation (4.49) requires formulation of boundary conditions at upper and lower surfaces, and sidewall, completed by symmetry conditions. The temperature distribution of a medium above the upper plate surface is not uniform. Assuming spherical symmetry of the temperature front propagation (Fig. 4.28a) the temperature distribution of the gas adjacent to the upper plate surface exhibits cylindrical symmetry

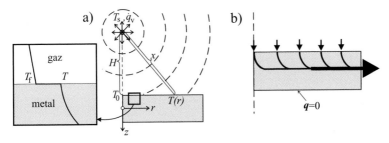

Fig. 4.28. Temperature front propagation (a) and schematic heat flow through plate (b)

and is given by the parabola $T(x) = T_s - \frac{\dot{q}_v}{6\lambda_g}x^2$, where T_s is the gas temperature at the heat source, λ_g denotes thermal conductivity of the gas, whereas x is a distance between the centre of the temperature front (heat source) and the particles of a gas neighbouring the current point of the plate. A typical heat flux through the plate is presented in Fig. 4.28b. It is assumed that the whole heat flux which enters the upper plate surface moves towards the sidewall, whereas the lower surface is subjected to adiabatic conditions. The above assumptions allow one to formulate mixed and von Neumann integral type boundary conditions for (4.49) in the following form:

$$\beta(T_f - T) = -\lambda\frac{\partial T}{\partial z} \quad \text{upper surface}$$
$$\partial T/\partial z = 0 \quad \text{lower surface}$$
$$\partial T/\partial r = 0 \quad \text{symmetry axis} \qquad (4.50)$$
$$\int_A -\lambda\frac{\partial T}{\partial r}\,\mathrm{d}A = C \quad \text{sidewall}$$

where β denotes the coefficient of Newtonian convection between the gas and the upper surface of the plate. The heat flux that enters the top surface of the plate and is sent through the sidewall is held constant throughout the process ($C = \text{const}$).

Equations of mechanical state of FGM

General formulation of *displacement equations of thermoelasticity* for FGM requires consideration of additional terms associated with derivatives of constitutive moduli with respect to coordinates (cf. Ganczarski [79])

$$\frac{1}{2}\text{div}\left[\mathbf{E} : (\nabla\mathbf{u} + \mathbf{u}\nabla)\right] = \text{div}\left(\mathbf{E} : \mathbf{1}\alpha\theta\right) \qquad (4.51)$$

If the functional gradation coincide with the transverse direction, then introducting cylindrical coordinates, (4.51) takes the following form:

$$\nabla^2 u - \frac{u}{r^2} + \frac{1}{1-2\nu}\frac{\partial \Theta}{\partial r} + \frac{1}{\mu}\frac{\partial \mu}{\partial z}\left(\frac{\partial u}{\partial z} + \frac{\partial w}{\partial r}\right) = 2\frac{1+\nu}{1-2\nu}\alpha\frac{\partial \theta}{\partial r}$$

$$\nabla^2 w + \frac{1}{1-2\nu}\frac{\partial \Theta}{\partial z} + \frac{2}{\mu}\frac{\partial \mu}{\partial z}\left(\frac{\partial w}{\partial z} + \frac{\nu\Theta}{1-2\nu}\right) = 2\frac{1+\nu}{1-2\nu}\alpha\left(\frac{1}{\mu}\frac{\partial \mu}{\partial z}\theta + \frac{\partial \theta}{\partial z}\right)$$

(4.52)

where dilatancy is given by $\Theta = \dfrac{\partial u}{\partial r} + \dfrac{u}{r} + \dfrac{\partial w}{\partial z}$, symbol θ denotes temperature difference, whereas additional terms enriching the problems are underlined. Similarly to the thermal problem, elastic modulus $\mu = E/(1+\nu)$ depends on the local volume fraction of ceramic in the Al-alloy and is modelled by Noda's approximation [166] (4.3) which, substituted into (4.52), leads to the final form of displacement equations:

$$\nabla^2 u - \frac{u}{r^2} + \frac{1}{1-2\nu}\frac{\partial \Theta}{\partial r} + k_2\left(\frac{\partial u}{\partial z} + \frac{\partial w}{\partial r}\right) = 2\frac{1+\nu}{1-2\nu}\alpha\frac{\partial \theta}{\partial r}$$

$$\nabla^2 w + \frac{1}{1-2\nu}\frac{\partial \Theta}{\partial z} + 2k_2\left(\frac{\partial w}{\partial z} + \frac{\nu}{1-2\nu}\Theta\right) = 2\frac{1+\nu}{1-2\nu}\alpha\left(k_2\theta + \frac{\partial \theta}{\partial z}\right)$$

(4.53)

Boundary conditions of the mechanical problem are of a more sophisticated nature than the conditions of the thermal problem. Simple support (Fig. 4.29a) is assumed along the sidewall of the plate under mechanical load p. This condition, however, constrains thermal transverse expansion at the sidewall, so, for the thermal problem a separate set of boundary conditions are proposed with point support at the lower corner (Fig. 4.29b). This trick requires decomposition of the displacement into purely mechanical and purely thermal components:

$$u = u^{\mathrm{m}} + u^{\mathrm{th}}$$
$$w = w^{\mathrm{m}} + w^{\mathrm{th}}$$

(4.54)

but allows one to formulate two sets of boundary conditions, which are used to solve the mechanical problem separately for the purely mechanical p and the purely thermal loading $\theta(r, z)$ (Table 4.8).

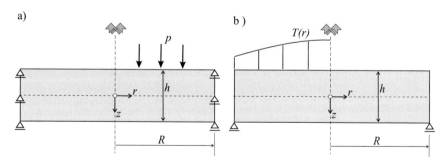

Fig. 4.29. Boundary conditions for purely mechanical load p (a), and purely thermal load $T(r)$ (b)

Table 4.8. Boundary conditions for purely mechnical and purely thermal loads

Mechanical state		Thermal state		Boundary
$\sigma_z = -p$	$\tau_{rz} = 0$	$\sigma_z = 0$	$\tau_{rz} = 0$	upper surface
$\sigma_z = 0$	$\tau_{rz} = 0$	$\sigma_z = 0$	$\tau_{rz} = 0$	lower surface
$u^m = 0$	$\partial w^m/\partial r = 0$	$u^{th} = 0$	$\partial w^{th}/\partial r = 0$	symmetry axis
$\sigma_z = 0$	$w^m = 0$	$\sigma_z = 0$	$\tau_{rz} = 0$	sidewall
		$w^{th} = 0$		lower corner

Numerical procedure

The *Finite Difference Method* (FDM) was developed as an approximate and discrete tool for solving differential boundary value problems. The essence of FDM is to replace differential operators by appropriate difference operators. Such an approximation of a function and its derivatives allows one to reduce the original boundary value problem to a system of algebraic equations of relatively simple form

$$\mathbf{A} \cdot \mathbf{x} = \mathbf{b} \tag{4.55}$$

In the case of the Fourier equation (4.49), the corresponding difference operator is shown in Fig. 4.30, whereas the boundary conditions allowing one to eliminate those of nodal unknowns situated outside the domain are given by formulas

$$
\begin{aligned}
T_{i,j-1} &= T_{i,j+1} + 2\Delta_z \frac{\beta}{\lambda} \left(T_{\mathrm{f}} - T_{i,j} \right) \text{ upper surface} \\
T_{i,j+1} &= T_{i,j-1} & \text{lower surface} \\
T_{i-1,j} &= T_{i+1,j} & \text{symmetry axis} \\
T_{i+1,j} &= T_{i-1,j} + 2\Delta_r g\left(z\right) & \text{sidewall}
\end{aligned}
\tag{4.56}
$$

Application of the *finite difference scheme* (4.30) to each node of a uniform

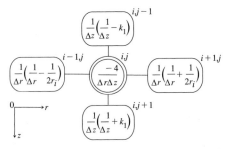

Fig. 4.30. Finite difference scheme of Fourier's equation (4.49)

mesh of size $n1 \times n2$ leads to the system of equations given by (4.55) of a band diagonal form. Namely, besides the main diagonal accompanied by one super- and sub-diagonal, forming together a tri-diagonal system, there are two additional diagonals situated symmetrically at distance $n2$ on both sides of bands, filled with zeros.

In order to obtain the finite difference representation of system (4.52) it is first transformed to the equivalent form

$$2\left(1-\nu\right)\left(\frac{\partial^2 u}{\partial r^2}+\frac{1}{r}\frac{\partial u}{\partial r}-\frac{u}{r^2}\right)+\left(1-2\nu\right)\left(\frac{\partial^2 u}{\partial z^2}+k_2\frac{\partial u}{\partial r}\right)+\frac{\partial^2 w}{\partial r\partial z}$$
$$+\left(1-2\nu\right)k_2\frac{\partial w}{\partial r}=2\left(1+\nu\right)\alpha\frac{\partial\theta}{\partial r}$$
$$\left(1-2\nu\right)\left(\frac{\partial^2 w}{\partial r^2}+\frac{1}{r}\frac{\partial w}{\partial r}\right)+2\left(1-\nu\right)\left(\frac{\partial^2 w}{\partial z^2}+k_2\frac{\partial w}{\partial z}\right)+\frac{\partial^2 u}{\partial r\partial z}+\frac{1}{r}\frac{\partial u}{\partial z}$$
$$+2k_2\nu\left(\frac{\partial u}{\partial r}+\frac{u}{r}\right)=2\left(1+\nu\right)\alpha\left(k_2\theta+\frac{\partial\theta}{\partial z}\right)$$

(4.57)

that allows for direct identification of differential operators

$$\mathcal{D}_1\left\{u,w\right\}=\mathcal{R}_1\left\{\theta\right\}$$
$$\mathcal{D}_2\left\{w,u\right\}=\mathcal{R}_2\left\{\theta\right\}$$

(4.58)

and then their replacement by corresponding finite difference schemes as shown in Figs 4.31 and 4.32. The finite difference representation of the bound-

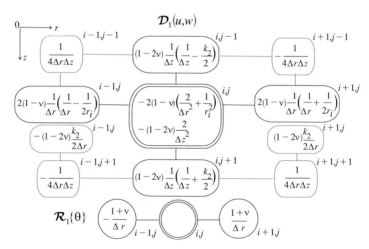

Fig. 4.31. Finite difference scheme of operators $\mathcal{D}_1$ and $\mathcal{R}_1$

ary conditions that allow one to eliminate nodal unknowns situated outside the domain is as follows

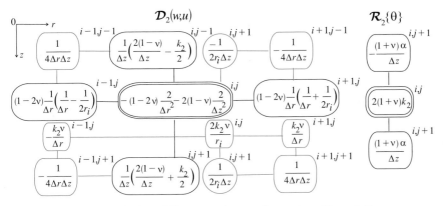

Fig. 4.32. Finite difference scheme of operators $\mathcal{D}_2$ and $\mathcal{R}_2$

$$
\left.
\begin{aligned}
&u_{i,j-1} = u_{i,j+1} - w_{i-1,j} + w_{i+1,j} \\
&w_{i,j-1} = w_{i,j+1} + \frac{\nu}{1-\nu}\left(-u_{i-1,j} + u_{i+1,j} + 2\frac{\Delta}{r_i}u_{i,j}\right) \\
&\qquad -2\Delta\frac{1+\nu}{1-\nu}\alpha\theta_{i,j} + 2\frac{1+\nu}{1-\nu}(1-2\nu)\Delta\frac{p}{E}
\end{aligned}
\right\}
\begin{array}{l}
\text{upper} \\
\text{surface}
\end{array}
$$

$$
\left.
\begin{aligned}
&u_{i,j+1} = u_{i,j-1} + w_{i-1,j} - w_{i+1,j} \\
&w_{i,j+1} = w_{i,j-1} - \frac{\nu}{1-\nu}\left(-u_{i-1,j} + u_{i+1,j} + 2\frac{\Delta}{r_i}u_{i,j}\right) \\
&\qquad +2\Delta\frac{1+\nu}{1-\nu}\alpha\theta_{i,j}
\end{aligned}
\right\}
\begin{array}{l}
\text{lower} \\
\text{surface}
\end{array}
\qquad (4.59)
$$

$$
\left.
\begin{aligned}
&u_{i,j} = 0 \\
&w_{i-1,j} = w_{i+1,j}
\end{aligned}
\right\}
\begin{array}{l}
\text{symmetry} \\
\text{axis}
\end{array}
$$

$$
\left.
\begin{aligned}
&u_{i+1,j} = u_{i-1,j} - 2\frac{\nu}{1-\nu}\frac{\Delta}{r_i}u_{i,j} + 2\Delta\frac{1+\nu}{1-\nu}\alpha\theta_{i,j} \\
&w_{i,j} = 0
\end{aligned}
\right\}
\begin{array}{l}
\text{sidewall}
\end{array}
$$

Again, application of the finite difference schemes (4.31) and (4.32) to each node of a uniform mesh of size $n1 \times n2$ requires introduction of an alternative system in which a sequence of difference equations in matrix $\mathbf{A}$ of size $2 \times n1 \times n2$ times $2 \times n1 \times n2$ corresponds with a sequence in the vector of unknowns $\mathbf{x}^\mathrm{T} = \{\dots, u_k, w_k, u_{k+1}, w_{k+1}, \dots\}$ and leads to a system of equations exhibiting multiband diagonal form. Namely, the main band of width 7 diagonals (main diagonal + 3 super-diagonals + 3 sub-diagonals) is accompanied by two additional bands of width 5 diagonals situated symmetrically at distance $2 \times n2$ on both sides of the main band.

It was shown that the finite difference representation of the thermoelastic problem leads to a system of N linear equations for N that reveals characteristic features of a sparse system, in which only a relatively small number of elements in matrix $\mathbf{A}$ is nonzero. In such a case traditional methods applied to tri-diagonal or band matrices, which may be stored in the operating memory of a computer in compact form, turns out to be completely ineffi-

cient. A natural way to avoid this inconvenience is the application of one of
the schemes based on indexed storage, which takes advantage of the fact that
only nonzero elements and additional information about their location in **A**
are stored, according to algorithm PCGPACK [179]. A scheme for row-indexed
sparse storage mode for difference operators $\mathcal{D}_{1,2}$ is shown in Fig. 4.33. The

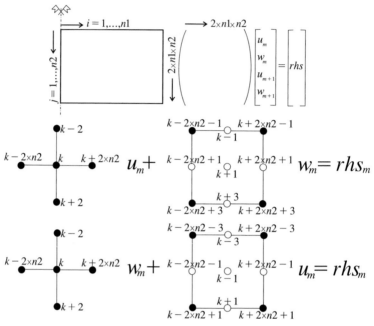

Fig. 4.33. Scheme of indexed storage for difference operators $\mathcal{D}_{1,2}$

conjugate gradient methods solves the $N \times N$ linear systems of form (4.55).
The attractiveness of these methods for large sparse systems is that the type
of operations on original matrix **A** are restricted to multiplication by a vector or perhaps transpose by a vector. One variant of the general algorithm
corresponding to the symmetric but nonpositive definite matrix **A** is called
minimum residual algorithm since it refers to successive minimizations of the
function

$$\Phi\left(\mathbf{x}\right) = \frac{1}{2}\mathbf{r} \cdot \mathbf{r} = \frac{1}{2}\left|\mathbf{A} \cdot \mathbf{x} - \mathbf{b}\right|^2 \tag{4.60}$$

the gradient of which

$$\nabla\Phi = \mathbf{A}^{\mathrm{T}} \cdot \left(\mathbf{A} \cdot \mathbf{x} - \mathbf{b}\right) \tag{4.61}$$

has a very convenient feature, namely that product $\mathbf{A}^{\mathrm{T}} \cdot \mathbf{A}$ is always symmetric
and positive definite for any matrix **A**. Rate of convergence of the conjugate
gradient methods depends on whether matrices are well-conditioned, in other
words "close" to the identity matrix. This suggests application of the so called

Preconditioned Biconjugate Gradient Method (PBCG) solving a system

$$\left(\widetilde{\mathbf{A}}^{-1} \cdot \mathbf{A}\right) \cdot \mathbf{x} = \widetilde{\mathbf{A}}^{-1} \cdot \mathbf{b} \tag{4.62}$$

instead of (4.55). Matrix $\widetilde{\mathbf{A}}$, termed a preconditioner, is close to $\mathbf{A}$ in the sense that $\widetilde{\mathbf{A}}^{-1} \cdot \mathbf{A} \approx \mathbf{1}$ and allows the algorithm to converge in a few steps. Numerical implementation of this algorithm is demonstrated in routine `linbcg.for` by Press [183], which requires double precision, since N is usually quite large.

Results

In what follows, the results of the engine piston analysis by the concept of FGM, is proposed after Cegielski [32]. A piston bottom having the following geometrical and loading characteristics is considered:

- thickness to radius ratio $h/R = 0.5$
- reference stress $\sigma_0 = 150$ MPa
- pressure of combustion gases over upper surface of piston $p = 10.0$ MPa
- temperature of combustion gases over centre of piston bottom $T_{f_0} = 300°C$
- reference temperature $T_{ref} = 20°C$
- coefficient of convection between combustion gases and upper surface of piston bottom $\beta = 100$ W/m^{2}K

Piston bottom made of homogeneous Al alloy

First let us consider a reference problem taking advantage of FDM code and results by Ganczarski [80]. The piston bottom is made of homogeneous Al alloy having the properties presented in Wang [232] and shown in Table 4.9.

Table 4.9. Material constants of Al alloy (after Wang [232])

Elastic mod. E [GPa]	Poisson's ratio ν	Coeff. ther. cond. λ [W/mK]	Coeff. ther. exp. α [1/K]
73	0.3	154	23×10^{-6}

Solution of the thermal problem decribed by (4.49) and boundary conditions (4.53) for homogeneous material $k_1 = 0$ gives the temperature distribution shown in Fig. 4.34.

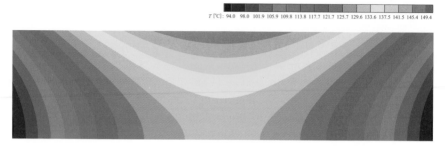

T [°C]: 94.0 98.0 101.9 105.9 109.8 113.8 117.7 121.7 125.7 129.6 133.6 137.5 141.5 145.4 149.4

Fig. 4.34. Temperature distribution in piston bottom

The highest temperature of 149.4°C appears on the top surface at a point on the to axis of symmetry, whereas the lowest temperature 94.0°C refers to a point located on the bottom surface adjacent to the sidewall. The heat flux field **grad**T corresponding to this temperature distribution is presented in Fig. 4.35. All local heat fluxes entering the upper surface of the piston

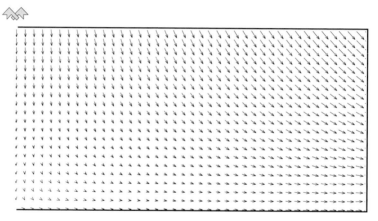

Fig. 4.35. Heat flux field **grad**T in piston bottom (mesh constant/vector length$=6.0 \times 10^{-3}$)

bottom turn to the sidewall, whereas both the axis of symmetry and lower surface satisfy adiabatic condition.

Solution of the mechanical problem given by (4.53) and boundary conditions (4.8) for homogeneous material $k_2 = 0$ turns out to be elementary in the case of the purely thermal state. Boundary conditions under consideration do not constrain freedom of thermal deformation and satisfy conditions necessary for the existence of a plane stress state according to the Sneddon theorem [212]. Therefore, taking into account that

$$\sigma_z = 2\mu \left[\frac{\partial w}{\partial z} - \alpha\theta + \frac{\nu}{1 - 2\nu} \left(\Theta - 3\alpha\theta \right) \right] = 0$$

$$\tau_{rz} = \mu \left(\frac{\partial w}{\partial r} + \frac{\partial u}{\partial z} \right)$$

(4.63)

and substituting into (4.53) we get an uncoupled system of equations

$$\nabla^2 u - \frac{u}{r^2} = (1 + \nu)\,\alpha\frac{\partial\theta}{\partial r}$$

$$\nu\nabla^2 w - \frac{\partial^2 w}{\partial z^2} = -(1 + \nu)\,\alpha\frac{\partial\theta}{\partial z}$$

(4.64)

in which the first equation has an elementary solution of Lamé type

$$u = (1 - \nu)\,\frac{\alpha r}{R^2} \int_0^R \theta r dr + (1 + \nu)\,\frac{\alpha}{r} \int_0^r \theta\xi d\xi$$

$$\sigma_r = E\alpha \left(\frac{1}{R^2} \int_0^R \theta r dr - \frac{1}{r^2} \int_0^r \theta\xi d\xi \right)$$

(4.65)

$$\sigma_\varphi = E\alpha \left(\frac{1}{R^2} \int_0^R \theta r dr + \frac{1}{r^2} \int_0^r \theta\xi d\xi - \theta \right)$$

whereas the second equation serves exclusively for establishing the axial displacement, since it refers to deformation not accompanied by stress. The corresponding finite difference scheme for the second part of (4.64) takes the form shown in Fig. 4.36. Deformation referring to the purely mechanical

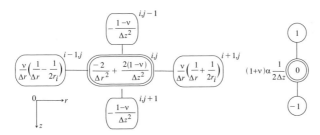

Fig. 4.36. Finite difference scheme for the second of (4.64)

state $\{u^m, w^m\}$ resulting from pressure of combustion gases p causes deflection downward, but simultaneously the influence of thermal field θ tends to cause deflection upward, is so strong that a distinct volume increase effect is observed $\{u^{th}, w^{th}\}$, such that total deformation is shown in Fig. 4.37. Both radial and hoop stress components are equal to -101.3 MPa at the point of maximum temperature (Fig. 4.38, Fig. 4.39), which together with compressive axial stress -10.0 MPa and zero shear stress means that r, φ, z are

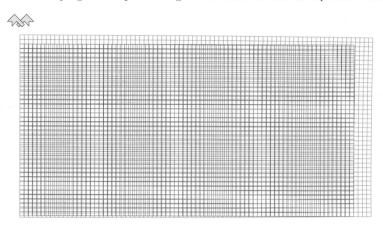

Fig. 4.37. Total deformation of piston bottom under pressure p and temperture θ (zoom $\times100$)

directions of eigenstress in the upper fibres. Hence stress concentration is advantageous since in the case of covering this surface with ceramic, compression will prevent cracking. Distribution of axial stress slightly differs both qualita-

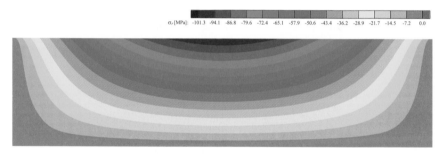

Fig. 4.38. Distribution of radial stress σ_r

tively and quantitatively from the classical distribution in Reissner's theory, and may be approximated with reasonable accuracy by the cubic parabola $-\frac{3}{4}p\left[\frac{2}{3} - \frac{2z}{h} + \frac{1}{3}\left(\frac{2z}{h}\right)^3\right]$ over almost the whole structure, except for a zone directly adjacent the sidewall (Fig. 4.40). Also shear stress may be properly approximated by the parabola $-\frac{3p}{4h}\left[1 - \left(\frac{2z}{h}\right)^2\right]$, because of its symmetry with regard to the midsurface, which is clearly visible in Fig. 4.41. Von Mises stress exhibits two maxima, the first equal to 85.6 MPa at a point located on the axis of symmetry on the upper surface and the other equal to 119.2 MPa at a point on the lower surface adjacent to the sidewall (Fig. 4.42).

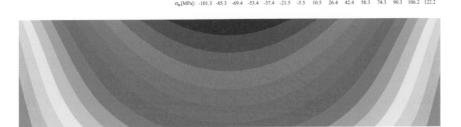

σ_φ[MPa]: -101.3 -85.3 -69.4 -53.4 -37.4 -21.5 -5.5 10.5 26.4 42.4 58.3 74.3 90.3 106.2 122.2

Fig. 4.39. Distribution of hoop stress σ_φ

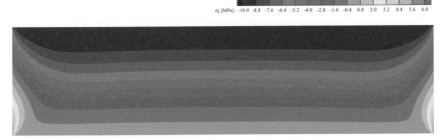

σ_z [MPa]: -10.0 -8.8 -7.6 -6.4 -5.2 -4.0 -2.8 -1.6 -0.4 0.8 2.0 3.2 4.4 5.6 6.8

Fig. 4.40. Distribution of axial stress σ_z

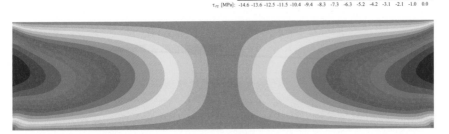

τ_{rz} [MPa]: -14.6 -13.6 -12.5 -11.5 -10.4 -9.4 -8.3 -7.3 -6.3 -5.2 -4.2 -3.1 -2.1 -1.0 0.0

Fig. 4.41. Distribution of shear stress τ_{rz}

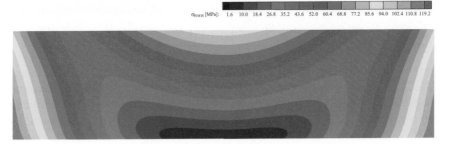

σ_{HMH} [MPa]: 1.6 10.0 18.4 26.8 35.2 43.6 52.0 60.4 68.8 77.2 85.6 94.0 102.4 110.8 119.2

Fig. 4.42. Distribution of von Mises stress σ_{HMH}

Piston bottom made of FGM composite Al/ZrO$_2$+Y$_2$O$_3$

For a piston bottom made of FGM composite Al/ZrO$_2$ stabilized by Y$_2$O$_3$, the thermomechanical properties of both components are shown in Table 4.10 (after Wang [232] and Lee [144]). The magnitudes of the coeffcients of material

Table 4.10. Material constants of Al/ZrO$_2$+Y$_2$O$_3$ (after Wang [232] and Lee [144])

Material	E [GPa]	ν	λ [W/mK]	α [1/K]
Al	73	0.3	154	23×10^{-6}
ZrO$_2$+Y$_2$O$_3$ (up to 1127°C)	205	–	2.0	9.8×10^{-6}

inhomogeneity defined by (4.3) are calculated from the following relations:

$$k_1 = \frac{2}{h} \ln \left(\frac{\lambda_{\mathrm{m}}}{\lambda_{\mathrm{c}}} \right) = 17.375, \qquad k_2 = \frac{2}{h} \ln \left(\frac{E_{\mathrm{m}}}{E_{\mathrm{c}}} \right) = -4.130 \qquad (4.66)$$

making the assumption that only fibres located over the midsurface are subjected to functional gradation.

Solution of the thermal problem described by (4.49) and boundary conditions (4.50) leads to temperature distribution shown in Fig. 4.43. In compar-

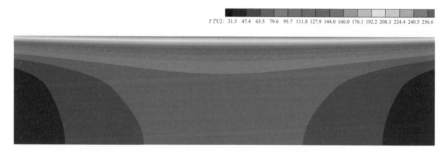

Fig. 4.43. Temperature distribution in piston bottom made of FGM

ison with temperature distribution for piston bottom made of homogeneous Al alloy, the new temperature distribution exhibits an increase of maximum temperature to 256.6°C and a simultaneous essential decrease of the minimum temperature to 31.3°C. As consequence of applying an FGM layer having a coefficient of thermal conductivity 77 times lower than the substrate, a strong insulation effect of the ceramic is observed, in which high temperature gradients appear and simultaneousely essential homogenization of the temperature field is observed in midsurface and lower fibres. The aforementioned effect is better visualized in Fig. 4.44 presenting the heat flux field. The maxium temperature gradient in the upper layers is increased almost 40 times in comparison with the reference solution. The solution of the mechanical problem

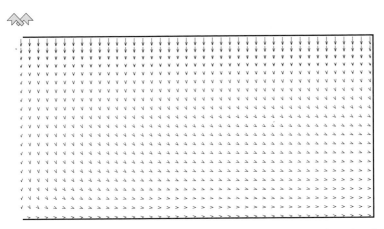

Fig. 4.44. Heat flux field **grad**T in piston bottom made of FGM (mesh constant/vector length=1.5×10^{-4})

given by (4.53), and boundary conditions (4.8), is illustrated by distributions of hoop stress, dominant in the analysis, shear stress and von Mises stress. Qualitative analysis of hoop stress shows essential differences in comparison with the reference solution (Fig. 4.45). Quantitative analysis of the distribu-

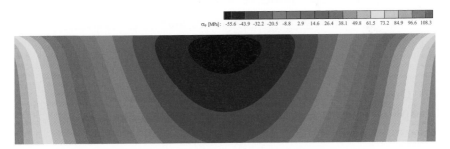

Fig. 4.45. Distribution of hoop stress σ_φ

tion confirms that application of the FGM system Al/ZrO_2+Y_2O_3, having a modulus of elasticity almost 2.8 times bigger than pure Al alloy, builds a thermal barrier and, as a consequence, a drop in the maximum hoop stress to -55.6 MPa at the central point of the top surface, and to 108.3 MPa on the bottom surface adjacent to the sidewall are noted. The above effect is accompanied by strong reduction of shear stress to -5.2 MPa on the bottom surface adjacent to the sidewall (Fig. 4.46) leading to an advantageous, from the strength capacity viewpoint, decrease of von Mises stress equal to 33.1

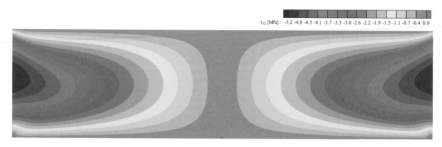

Fig. 4.46. Distribution of shear stress τ_{rz}

MPa at the central point of the top surface and 108.0 MPa on the bottom surface adjacent to the sidewall, respectively.

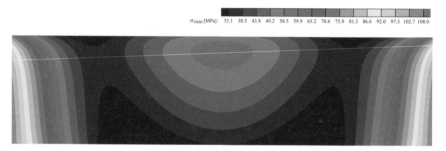

Fig. 4.47. Distribution of von Mises stress σ_{HMH}

4.3 Developing and Implementing Models of the Coupled Thermo–elasto–plastic–damage FGMs for Thermomechanical Applications

4.3.1 Extension of the Thermo–elasto–plastic–damage Model by the FGMs Concept

Definition of composite materials

Two kinds of composite materials are used in this section in order to implement the concept of functionally graded material to simple plane structural FGM and TBC systems suffering from damage and plasticity under mechanical or thermal loadings.

Metal–metal (MM) composite

Metal–metal (MM) composite is used for simulation of the structural response of a FGM-system under mechanical loading.

Two grades of spheroidized graphite cast iron materials P_1 and P_2 are used as constituents of the composite material (Table 4.11). Values of the model constants refer to the elasto–plastic–damage material described in Sect. 3.2. In what follows it was assumed that the mechanical nonhomogeneities are

Table 4.11. Material properties (mechanical) of two grades of spheriodized graphite cast iron used for FGMs simulation (after Egner *et al.* [65])

	E [MPa]	ν [-]	ϑ_1 [MPa^{-1}]	ϑ_2 [MPa^{-1}]
P_1 $\xi = 0$	169×10^3	0.285	-3.95×10^{-7}	4.00×10^{-6}
P_2 $\xi = 1$	85×10^3	0.285	-3.97×10^{-7}	4.00×10^{-6}
	ϑ_3 [MPa^{-1}]	ϑ_4 [MPa^{-1}]	ζ [-]	K_d [MPa]
P_1 $\xi = 0$	-4.00×10^{-7}	2.50×10^{-6}	0.89	1.3
P_2 $\xi = 1$	-4.00×10^{-7}	2.50×10^{-6}	0.89	1.3
	b [-]	R_0 [MPa]	R_∞ [MPa]	B_0 [MPa]
P_1 $\xi = 0$	15	293.0	250.0	0.273
P_2 $\xi = 1$	15	250.0	200.0	0.273
	c^P [-]	c^d [-]	c^r [-]	ρ [kgm^{-3}]
P_1 $\xi = 0$	1.0	-15.0	50.0	7840
P_2 $\xi = 1$	1.0	-15.0	50.0	7840

introduced in such a way that the volume fraction of graphite particles in the ferritic matrix varies in one direction, causing smooth change of the elasticity modulus E and yield stresses, initial R_0 and asymptotic R_∞. The sensitivity of the constitutive model (3.56) through (3.69) to changes of other parameters was also checked but was found to be less important, hence, for simplicity, it is ignored and the same Hayakawa–Murakami calibration (Table 3.1) is used.

A simple power-law continuous unidirectional gradation of constituent properties is used (Naghdabadi and Hosseini Kordkheili [164])

$$P(\xi) = (P_2 - P_1)\xi^n + P_1 \tag{4.67}$$

where n is the power-law exponent that indicates the degree of material grading, and ξ is the local coordinate of the sample width, referred to the sample thickness, $\xi = z/h(x)$ (Fig. 4.48).

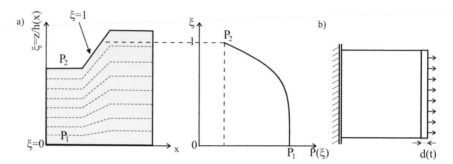

Fig. 4.48. (a) FGM system for unidirectional gradation, (b) rectangular sample for a homogeneous uniform test

In order to illustrate the material response of homogeneous composite materials of different composition (as characterized by a number ξ), a plane rectangular sample of constant thickness, fixed at one edge and uniformly loaded at the free edge (Fig. 4.48b), was analysed. The variable effective material properties used for a simulation are given in Table 4.12.

Table 4.12. Effective material properties used for simulation in the homogeneous, uniform test ($n = 5$)

No	ξ	% FCD 400	E[MPa]	R_0[MPa]	R_∞[MPa]	$R_0 + R_\infty$[MPa]
1	0.00	100	169 000	293.0	250.0	543.0
2	0.70	83.2	154 882	285.8	241.6	527.4
3	0.80	67.2	141 475	278.9	233.6	512.5
4	0.90	41.0	119 399	267.6	220.5	488.1
5	0.95	22.6	104 002	259.7	211.3	471.0
6	1.00	0.0	85 000	250.0	200.0	450.0

The uniform, uniaxial stress–strain characteristics of different homogeneous materials under tension and compression, versus the volume fraction parameter ξ, are presented in Fig. 4.49a, b.

Note that the more brittle the composite material used (lower values of ξ parameter) the elastic-damage phase, which precedes plastic flow, becomes more pronounced. If the ductility of a material is sufficiently high (higher values of ξ parameter), the elastic-damage phase can be avoided due to a prior

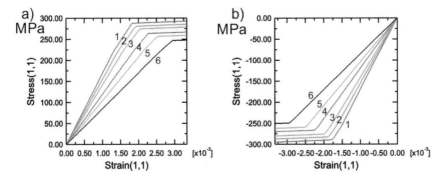

Fig. 4.49. Uniaxial tensile and compressive stress–strain curves for homogeneous elasto-plastic-damage spheroidized graphite cast iron having various fractions of graphite and metallic phases: (a) tension, (b) compression

plastic softening of the material. Roughly speaking, plasticity- and damage-based dissipative mechanisms interact with each other in a competitive fashion – the more ductility in the material response, the higher the resistance of the material to elastic damage. Nevertheless, plastic strains may be dangerous and have to be controlled.

Different tension/compression behaviour of the materials is not obvious in stress–strain curves due to a relatively weak uniaxial effect assumed for FCD 400 material grades ($\xi = 0.89$, cf. (3.54)). However, damage growth characteristics differ significantly in tensile tests and compressive tests (Fig. 4.50a, b)

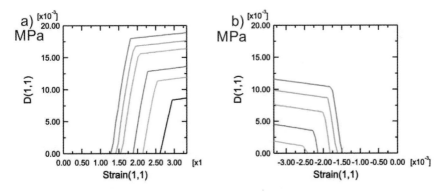

Fig. 4.50. Simulation of unilateral damage response under tension (a) and compression (b)

The material resistance to damage under compression is clearly higher than under tension, both in the sense of corresponding lower damage values and longer times for damage initiation. Note that for grade No. 6 the material damage does not occur under the compressive edge displacement range of $d = 4$ mm used for the simulation.

Metal–ceramic FCD400/ZrO$_2$ composite

Metal–ceramic (MC) composite is used for the simulation response of a TBC–FGM–substrate system under thermal loading. Two constituent materials are used to define a thermally resistant system composed of: TBC (zirconium oxide ZrO$_2$)–FGM–substrate (cast iron).

The mechanical properties of the metallic substrate are as for the spheroid-ized graphite cast iron FCD400 at room temperature detailed by Hayakawa and Murakami [98] in Table 4.13. For the thermal coating (TBC–ZrO$_2$), basic

Table 4.13. Mechanical and thermal properties of cast iron FCD400 at room temperature (material P_1)

E[MPa]	ν[–]	ϑ_1[MPa^{-1}]	ϑ_2 [MPa^{-1}]	ϑ_3[MPa^{-1}]	ϑ_4[MPa^{-1}]	ζ[–]
169×10^3	0.285	-3.95×10^{-7}	4.00×10^{-6}	-4.00×10^{-7}	2.50×10^{-6}	0.89
K_d[MPa]	b[–]	R_0[MPa]	R_∞[MPa]	B_0[MPa]	c^P[–]	c^d[–]
1.3	15	293.0	250.0	0.273	1.0	-15.0
c^r[–]	ρ[kgm^{-3}]	$\alpha[10^{-6}/K]$	λ[W/mK]	c[J/kgK]		
50	7840	11.4	83.5	480		

mechanical properties are shown in Table 4.14, (Egner *et al.* [65]). Additional

Table 4.14. Mechanical and thermal properties of zirconium oxide at room temperature (material P_2)

E[MPa]	ν[–]	ϑ_1[MPa^{-1}]	ϑ_2 [MPa^{-1}]	ϑ_3[MPa^{-1}]	ϑ_4[MPa^{-1}]	ζ[–]
116.4×10^3	0.25	4.5×10^{-6}	-2.04×10^{-6}	-2.23×10^{-6}	5.36×10^{-5}	0.1
K_d[MPa]	b[–]	R_0[MPa]	R_∞[MPa]	B_0[MPa]	c^P[–]	c^d[–]
6.979	15	–	–	2.78	0.0	0.0
c^r[–]	ρ[kgm^{-3}]	$\alpha[10^{-6}/K]$	λ[W/mK]	c[J/kgK]		
0.0	5560	13.3	1.71	274		

thermal properties of both constituents at room temperature are taken from Eshbach and Souders [67] (metal), and Ootao *et al.* [173] (ceramic). Note that in the zirconia constituent P_2, plastic behaviour was ignored for simplic-ity. Mechanical and thermal properties of both P_1–cast iron and P_2–zirconium oxide are temperature dependent. In the case of a nonuniform temperature field the resulting material thermally induced nonhomogeneity has to be accounted for. Temperature dependence of the basic material properties of metal are sketched in Fig. 4.51, (Ottosen and Ristinmaa [176]). Other material data

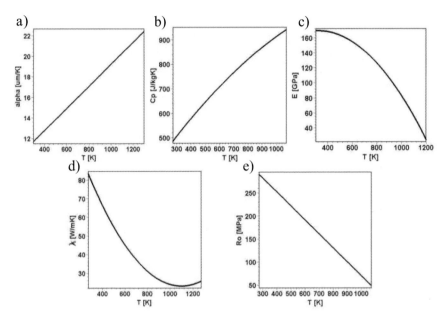

Fig. 4.51. Temperature dependence of principal mechanical and thermal properties of metal: (a) coefficient of thermal expansion α; (b) specific heat c; (c) Young modulus E; (d) thermal conductivity λ; (e) yield stress R_0

not presented in Fig. 4.51 are, in what follows, not altered with temperature change (no data available), however, their influence on the structural response is less important.

For the zirconium oxide ceramic temperature dependence of material properties is given by (1.1). For a continuous properties gradation through the thickness of the FGM-layer and temperature dependent properties of both constituents $P_1(T)$ and $P_2(T)$, the following approximation is used

$$P_1(\xi, T) = \frac{1}{2}\left[P_1(T) - P_2(T)\right]\sin[\pi(250\xi - 8.75)] + \frac{1}{2}\left[P_1(T) + P_2(T)\right] \quad (4.68)$$

where ξ is the local coordinate of the sample width (in the FG-direction) referred to the sample thickness (Fig. 4.48) and functions $P_1(T)$ and $P_2(T)$ denote variation with temperature of chosen properties of materials entering into the composition of FGM (FCD400–P_1 and ZrO_2–P_2 respectively).

4.3.2 Prediction of Damage in the Plane-stress Notched Specimen by use of FGMs Under Mechanical Loading

A double-notched plane MM composite specimen, clamped at one edge and free to move at the other, is subjected to monotonously increasing, uniformly distributed edge displacement $d(t)$ (Egner *et al.* [65]). A more ductile (softer) spheroidized graphite cast iron material P_2 is deposited at the notched edge of the sample, and smooth properties gradation from that of the edge coating to the interior (P_1) is assumed according to (4.67), cf. Fig. 4.48a and Table 4.12. Geometry of the sample and FE mesh are shown in Fig. 4.52.

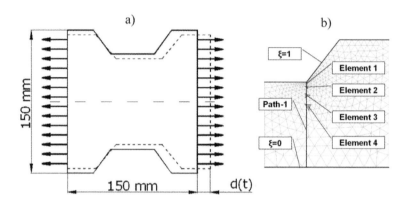

Fig. 4.52. A plane-stress double-edge notched sample subjected to uniform tension: (a) geometry and load; (b) FE mesh, after Egner *et al.* [65]

The response of the structure to uniformly increasing tensile load is simulated numerically for two cases: (A) sample is made of a homogeneous, reference material (HM) P_1 (cf. Table 4.11), and (B) a functionally graded sample is made of a composite of P_1 and P_2 constituents, the properties of which change smoothly from P_1 ($\xi = 0$) in the interior to P_2 ($\xi = 1$) at the notched edge (FGM) (cf. 4.67). Note that, for fabrication reasons, a layered structure would be used, hence the smooth properties gradation used in this example is to be considered as an approximation only.

A comparison of the HM versus the FGM response to an increasing uniform edge displacement up to $d = 0.4$ mm, shows prospective advantages of the application of a functionally graded structure. Due the "softer" material used at the notched edge compared to the "stronger" material in the interior, mitigation of the stress concentration in the vicinity of the notch is observed (Fig. 4.53). It was found that, for the ultimate edge displacement $d = 0.4$ mm, the maximum principal stress at the notch (Point 1) in the homogeneous sample equals 472.1 MPa, compared with the corresponding value of 367.6 MPa (reduced by 22%) in the FGM sample. Additionally, a more balanced stress distribution across the width at the notch in the FGM-structure when

a)

b)

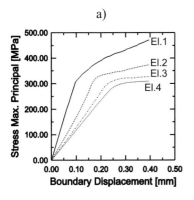
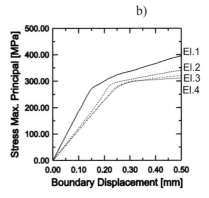

Fig. 4.53. Maximum principal stress versus edge displacement: (a) homogeneous material HM, (b) functionally graded material FGM

compared to that in the HM-structure was obtained. Application of the functionally graded material causes reduction in the stress concentration at the notch, as shown in Fig. 4.54.

In the case of homogeneous material (HM) a spurious stress concentration at the notch results in earlier damage initiation when compared to the functionally graded material (FGM). Application of FGM retarded microcrack initiation at the notch (El. 1) measured by the critical edge displacement, from $d = 0.07$ mm for HM to $d = 0.12$ mm for FGM. At the point at the interior (El. 4) the activation of damage occurs for similar values of edge displacement (Fig. 4.55).

Maximum values of damage components D_{11} at El. 1 for the edge displacement $d = 0.40$ mm are also reduced from $D_{11} = 0.03371$ for HM to $D_{11} = 0.02998$ for FGM, and the damage distribution is more balanced when FGM is used. A similar effect is observed for the decreased scalar damage hardening variable β (a measure of cumulative damage) in the FGM structure when compared to the homogeneous one (Fig. 4.56).

Damage spatial progress around the notch is more balanced when the FG structure is tested (Fig. 4.58a) compared to the uniform structure (Fig. 4.58b). However, the inclined damage zone propagating inwards from the notch foreshadows the coming failure in both cases.

Some reduction of plastic strains in a functionally graded sample when compared to a homogeneous one is also observed, both for the plastic strain concept ε_{11}^{P} and the cumulative scalar damage variable r (Fig. 4.57).

Advanteges of the use of functionally graded double-notched sample versus the homogeneous one are summarized in Table 4.15, compared for the ultimate edge displacement $d = 0.40$ mm.

a)

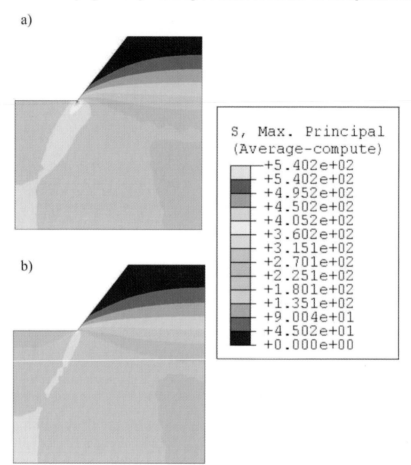

b)

```
S, Max. Principal
(Average-compute)
 ─┌── +5.402e+02
  ├─ +5.402e+02
  ├─ +4.952e+02
  ├─ +4.502e+02
  ├─ +4.052e+02
  ├─ +3.602e+02
  ├─ +3.151e+02
  ├─ +2.701e+02
  ├─ +2.251e+02
  ├─ +1.801e+02
  ├─ +1.351e+02
  ├─ +9.004e+01
  ├─ +4.502e+01
 ─└── +0.000e+00
```

Fig. 4.54. Distribution of the maximum principal stress at load stage $d = 0.4$ mm: (a) in homogeneous material HM, (b) in functionally graded material FGM

Table 4.15. Comparison of results at notch (El. 1) for: σ_{max}, ε_{max}, D_{11}, ε_{11}^{P}

Sample material	σ_{max} [MPa]	ε_{max} [–]	D_{11} [–]	ε_{11}^{P} [–]
Homogeneous	472.1	0.00748	0.03371	0.05223
Functionally graded (E, R_0, B_0)	367.6	0.00859	0.02998	0.03293

A significant reduction of stress, damage and plastic strain at the notch is clear, although the maximum total strain for the FGM-structure is higher than that for the HM-structure. Besides the reduction in magnitude of ultimate damage and plastic strain ($d = 0.40$ mm), the initiation of both in El. 1

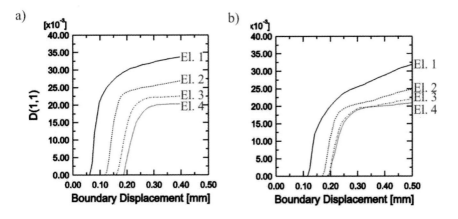

Fig. 4.55. Principal component of damage tensor D_{11} versus the edge displacement: (a) homogeneous material HM, (b) functionally graded material FGM

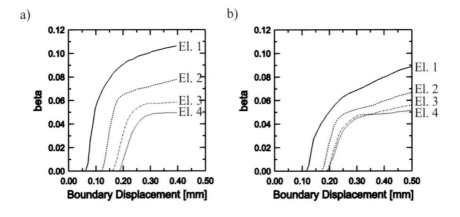

Fig. 4.56. Scalar damage variable β versus the edge displacement: (a) homogeneous material HM, (b) functionally graded material FGM

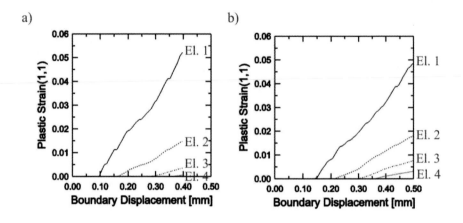

Fig. 4.57. Plastic strain $\varepsilon_{11}^{\mathrm{P}}$ versus edge displacement: (a) homogeneous material HM, (b) functionally graded material FGM

is retarded in the FGM-system, as shown in Table 4.16. However, in El. 4 (the interior), the incubation time for both dissipative phenomena are slightly accelerated.

Table 4.16. Comparison of damage and plasticity incubation periods at notch (El. 1) and interior (El. 4)

Sample material	$d_{\mathrm{inc}}(D_{11})$ [mm]		$d_{\mathrm{inc}}(\varepsilon_{11}^{\mathrm{P}})$ [mm]	
	El. 1	El. 4	El. 1	El. 4
Homogeneous	0.07	0.19	0.10	0.36
Functionally graded (E, R_0, B_0)	0.12	0.19	0.15	0.34

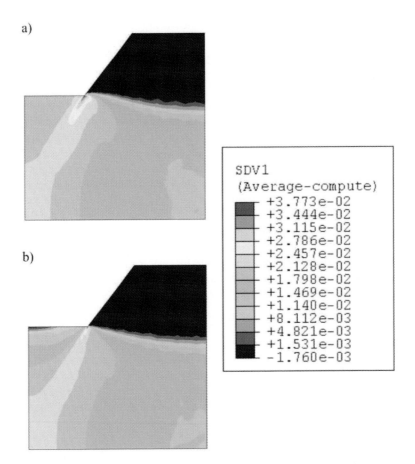

Fig. 4.58. The distribution of damage component D_{11} at load stage $d = 0.4$ mm: (a) in homogeneous material HM, (b) in functionally graded material FGM

4.3.3 Analysis of a Problem with the ZrO_2/FCD400 TBC/FGM/S System Under Thermal Loading

In what follows the constitutive model of a *thermo–elasto–plastic–damage material*, developed in Sect. 3.2, is used to simulate a complex thermo-insulating system, TBC–FGM–substrate, subjected to thermal loading. A composite material consists of two constituents: P_1 – spheroidized graphite cast iron FCD400 (Hayakawa and Murakami [98]) and P_2 – zirconium oxide ZrO_2 (Egner *et al.* [65]), as described in Table 4.13 (P_1) and Table 4.14 (P_2). Additional temperature dependence of the principal material data is shown in Fig. 4.51 (cast iron), while for ceramic it is given by (1.1).

The plane-stress structure considered is shown in Fig. 4.59a. Three systems are examined; (a) reference homogeneous metal (FCD400) structure, (b) two-layer structure that consists of a 2 mm thick TBC (ZrO_2) and a metal (FCD400) substrate, and (c) three-layer system (TBC/FGM/substrate): 2 mm thick thermal barrier coating of zirconium oxide bonded with the substrate metallic material by a 4 mm thick layer of FGM. In all cases the geometry, boundary conditions and loading of the structure remain the same. Due to the symmetry, only a half of the structure was used for computation. FE mesh contains 1208 quadrilateral elements, with thermal degrees of freedom, nonuniformly distributed, as shown in Fig. 4.59b.

Two steps are applied in order to simulate complete loading cycle: phase 1 - uniform cooling of the structure from the elevated fabrication temperature T_1=600 [K] to the reference temperature T_{ref}=300 [K] during the time period t_1=180 [s] (the fabrication phase) and then, phase 2 - nonstationary heat flux of intensity $Q = 5 \times 10^5$ [J/(m^2 s)], uniformly applied at the top (thermal layer, Fig. 4.59a) and propagating into the metallic substrate during the period t_2=40 [s] (the working phase). Fabrication phase was simulated by a uniform, linear drop of temperature throughout the structure, with both sides, left and right, kept free to move. Due to nonhomogeneity, residual stresses are built in the composite FG-structure. During the heating phase 2, horizontal displacements were ruled out from both sides, left- and right-hand (see Fig. 4.59a).

In the homogeneous structure no residual stress appears (Fig. 4.60a). In the case of the two-layer system a disadvantageous jump of the residual stress occurs due to rapid properties mismatch at the TBC/substrate interface (Fig. 4.60b). In the three-layer system residual stress changes "smoothly" through the FGM layer (Fig. 4.60c).

When the three-component system TB/FGM/substrate is used, a smooth temperature variation, from that of the coating material to the substrate is observed (Fig. 4.61c). Such a temperature change is more promising, since it results in a lower stress mismatch at the interface, and hence, failure at the coating may be avoided. It is observed when the ultimate stress distributions are compared at the end of the phase 2 (Fig. 4.62a-c).

Inspection of plastic strain growth in structures shows benefits of using the TBC/FGM/substrate system when compared to the other systems. It is seen in Fig. 4.63, where scalar plastic hardening variable r distribution is shown. The plastic strain growth in the three-layer structure is limited to the small zone near the left-hand-side boundary, in the substrate material neighbouring the FG-layer as shown in Fig. 4.63c, whereas both the reference and two-layer structures suffer from plastic strain development in the metallic material, Figs 4.63a and 4.63b.

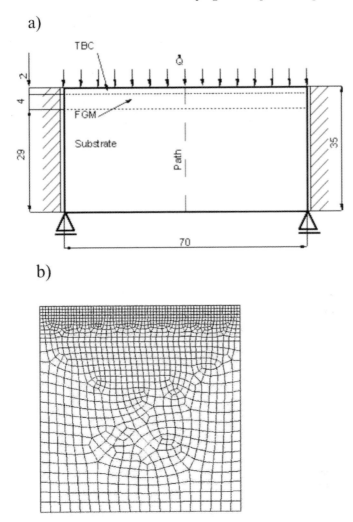

Fig. 4.59. TBC/FGM/substrate system: (a) geometry and loading, (b) FE mesh

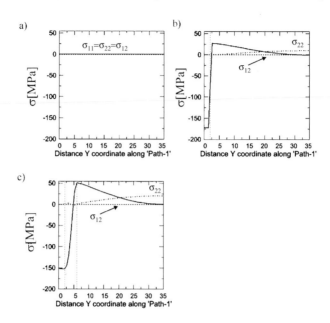

Fig. 4.60. Residual stress distribution along the path (Fig. 4.59a) at the end of the cooling phase

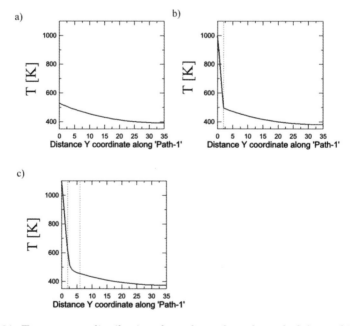

Fig. 4.61. Temperature distribution along the path at the end of the working phase

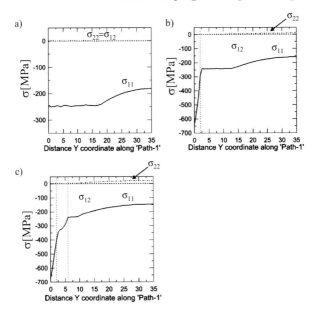

Fig. 4.62. Comparison of the thermally induced ultimate stresses along the path at the end of the working phase

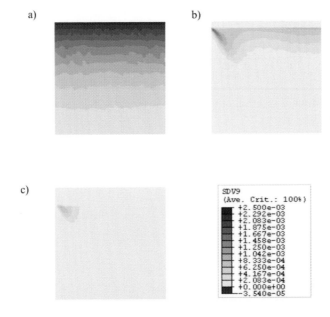

Fig. 4.63. 2-D distribution of the cumulative plastic strain r through the volume at the end of the working phase

Moreover, when damage incubation periods are considered at points of maximum value of scalar damage variable, the three-layer structure is also most beneficial, as shown in Fig. 4.64, where β - evolution in time is visualized.

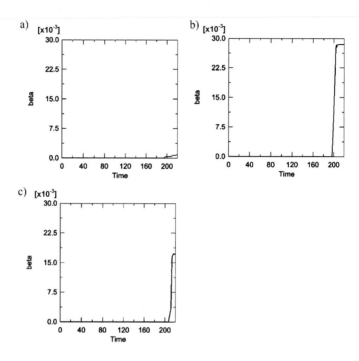

Fig. 4.64. Evolution of variable β at points of highest damage level

5

Microstructural Analysis and Residual Stress Determination based on Scattering of Neutrons and X-ray Synchrotron Radiation

5.1 Introduction to the Scattering of X-rays and Neutrons

A detailed treatment of the subjects of this section can be found in Batterman and Cole [16], Brückel [26], Cantor and Schimmel [30], Cowley [52], Dianoux and Lander[55], Feigin and Svergun [74], Glatter and Kratky [88], Guinier and Fournet [90], Jackson [117], Jacrot [118], Landau and Lifshitz [140], Sears [200], Squires [214].

5.1.1 Unperturbed beams

X-rays

X-rays are electromagnetic waves with a typical range of wavelength λ between 0.05 and 0.2 nm. Let us consider a monochromatic X-ray wave propagating in vacuum. From the classical point of view, it is known that it is constituted by oscillations of the electric field vector, $\mathbf{E}$, and of the magnetic field vector, $\mathbf{B}$, over perpendicular planes, which are both perpendicular to the direction of wave propagation. Both $\mathbf{E}$ and $\mathbf{B}$ can be obtained by *d'Alembert's equation*

$$\nabla^2 \mathbf{E} = -\boldsymbol{\nabla} \times \boldsymbol{\nabla} \times \mathbf{E} = \frac{1}{c^2} \frac{\partial \mathbf{E}}{\partial t} \tag{5.1}$$

where $\boldsymbol{\nabla}$ is the gradient operator and c is the speed of light in vacuum. The general solution is

$$\mathbf{E}(\mathbf{r}, t) = \mathbf{E}_0 \, e^{\imath(\mathbf{k}_0 \cdot \mathbf{r} - \omega t)} \tag{5.2}$$

where $\mathbf{k}_0$ is the so-called wave vector, with modulus $k_0 = 2\pi/\lambda$ and collinear with the direction of wave propagation, ω is the angular frequency, $\omega = 2\pi\nu$, ν being the wave frequency, and $\imath = \sqrt{-1}$ (Fig. 5.1). Notice that the relation $c = \omega/k_0$ holds for any reference frame. It is also known that electromagnetic waves have a corpuscular nature: the transported energy is quantized in bundles

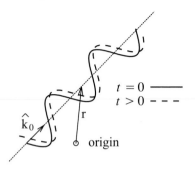

Fig. 5.1. Electrical component of a monochromatic polarized electromagnetic wave

(called photons) of energy $\mathcal{E} = h\nu = \hbar\omega$, where h is Planck's constant ($\hbar = h/2\pi$).

Neutrons

Let us consider now a neutron beam. These particles, provided with a mass m_N and a spin $I_N = \frac{1}{2}$, are described in the framework of material wave mechanics. In particular, if their speed is much lower than c, the nonrelativistic quantum mechanics theory can be adopted. The material wave function $\Psi(\mathbf{r}, t)$ is a solution of the *time-dependent Schrödinger equation*

$$\widehat{H}\Psi = \imath\hbar\frac{\partial\Psi}{\partial t} \tag{5.3}$$

where $\widehat{H}$ is the Hamiltonian operator

$$\widehat{H} = -\frac{\hbar^2}{2m_N}\nabla^2 + U \tag{5.4}$$

with U the potential energy. In particular, if U does not depend on time t, the wave function can be factorized into a product of a *space-wave function* $\psi(\mathbf{r})$ and a *time-wave function* $\phi(t)$. The space-wave function is the solution of the *time-independent Schrödinger equation*

$$\widehat{H}\psi = \mathcal{E}\psi \tag{5.5}$$

where $\mathcal{E}$ is the energy associated with the state $\psi(\mathbf{r})$. The time-wave function is the straightforward solution of

$$\imath\hbar\frac{\partial\phi}{\partial t} = \mathcal{E}\phi \quad \Leftrightarrow \quad \phi(t) = e^{-\imath\frac{\mathcal{E}}{\hbar}t} \tag{5.6}$$

For a free neutron beam ($U = 0$), the simple solution of (5.5) is

$$\psi(\mathbf{r}) = \frac{1}{L^{3/2}}e^{\imath\mathbf{k}_0\cdot\mathbf{r}} \tag{5.7}$$

where the wave vector $\mathbf{k}_0$ represents the momentum of the neutron in $\hbar$ units, with modulus $k_0 = (2m_N\mathcal{E})^{1/2}/\hbar$. L, which appears in the normalization constant, represents the side of the box that will contain the portion of the sample that will be irradiated by the neutron beam. The whole wave function is then

$$\Psi(\mathbf{r}, t) = \frac{1}{L^{3/2}} e^{\imath(\mathbf{k}_0 \cdot \mathbf{r} - \frac{\mathcal{E}}{\hbar}t)} \tag{5.8}$$

It is important to note the resemblance of this equation with (5.2): both relations represent plane stationary waves. For neutrons, Ψ is a scalar field; for X-rays $\mathbf{E}$ is a vectorial field. Moreover, for neutrons the energy and the wavelength are given by the classical kinetic energy and by *de Broglie's equation*

$$\mathcal{E} = \frac{1}{2}m_N v_N{}^2 = \frac{\hbar^2 k_0^2}{2m_N} \tag{5.9}$$

$$\lambda = \frac{h}{m_N v_N} = \frac{2\pi}{k_0} \tag{5.10}$$

with v_N the neutron velocity. Typical λ values for nonrelativistic neutrons (also called cold or thermal neutrons) are between 0.1 and 1 nm.

5.1.2 Interactions

X-rays

Let us consider now the interaction between X-rays and matter. Due to the their electromagnetic nature, X-rays will essentially interact with charged particles. For the sake of simplicity, we consider in detail the classical elastic interaction with an electron. This electron belongs to a material system and we suppose that, classically, it has its own (natural) angular frequency $\omega_0 \equiv (\kappa/m)^{1/2}$, with κ the elastic force constant and m the electron mass. The average position of the electron is conveniently chosen as the origin of the reference frame (Fig. 5.2). Moreover, we consider a polarized X-ray beam that propagates along the z axis with its electric component $\mathbf{E}$ oscillating on the xz plane. In this case, from (5.2), we have

$$E_x = E_0\, e^{\imath(k_0 z - \omega t)} \tag{5.11}$$

The electric field that acts on the electron at the origin $(z = 0)$ is $E_x = E_0\, e^{-\imath\omega t}$ and the corresponding electric force will be $F_x = -eE_x = -eE_0\, e^{-\imath\omega t}$, with $-e$ the charge of the electron. *Newton's equation of motion* will contain the electric force and the elastic (Hook) force,

$$m\frac{\mathrm{d}^2 x}{\mathrm{d}t^2} = -eE_0\, e^{-\imath\omega t} - m\omega_0^2 x \tag{5.12}$$

The general solution of the equation of motion is

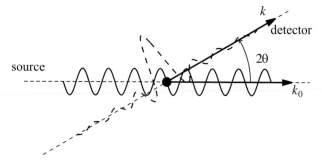

Fig. 5.2. Elementary process of scattering. The scattering particle becomes the source of a spherical wave

$$x(t) = -E_0\, e^{-\imath\omega t}\, \frac{e}{m}\, \frac{1}{\omega_0^2 - \omega^2} \tag{5.13}$$

and the corresponding acceleration is

$$a_x(t) = \frac{d^2 x}{dt^2} = E_0\, e^{-\imath\omega t}\, \frac{e}{m}\, \frac{\omega^2}{\omega_0^2 - \omega^2} \tag{5.14}$$

Subjected to an acceleration, the electron will irradiate its own electromagnetic wave. The electric component of this wave at the position $\mathbf{r}$ (where we will put the detector, Fig. 5.2) and at the time t can be calculated from the general expression

$$\mathbf{E}(\mathbf{r}, t) = -\frac{e}{c^2 r^3}\mathbf{r} \times [\mathbf{r} \times \mathbf{a}(t - r/c)] \tag{5.15}$$

which, for our particular system, leads to the following expression for the modulus of $\mathbf{E}(\mathbf{r}, t)$:

$$E(\mathbf{r}, t) = E_0\, e^{-\imath\omega t}\, \frac{e^2}{mc^2}\, \frac{\omega^2}{\omega_0^2 - \omega^2}\, \sin\alpha\, \frac{e^{\imath k_0 r}}{r} \tag{5.16}$$

with α the angle between $\mathbf{r}$ and the x axis. This is the classical expression for *Rayleigh scattering*. The first factor $r_0 = \frac{e^2}{mc^2}$ has dimension of a length and it is called the classical radius of the electron, $r_0 = 0.28 \times 10^{-12}$ cm. The second factor is the frequency factor. Two cases can be distinguished. If $\omega_0 \gg \omega$, as for visible light, the lower is ω_0 the higher is the frequency factor. This is the blue-shift phenomenon, responsible, for example, for blue sky. On the other hand, when $\omega_0 \ll \omega$ (as occurs essentially for X-rays) the frequency factor is -1. This is the case of *Thompson scattering*:

$$E(\mathbf{r}, t) = -E_0 e^{-\imath\omega t}\, \frac{e^{\imath k_0 r}}{r}\, r_0 \sin\alpha \tag{5.17}$$

The angular factor $\sin\alpha$ indicates that the maximum amplitude occurs for $\alpha = \pi/2$. Finally, and very important, the factor $\frac{e^{\imath k_0 r}}{r}$ represents a spherical wave. Thus the electron becomes the source of a spherical wave, with the same frequency (and λ) as the striking X-ray. The incident power per unit surface (i.e. the beam intensity) I_0 is represented by the time-average modulus of the *Poynting vector* $\mathbf{S} \equiv \frac{c}{4\pi}\mathbf{E} \times \mathbf{B}$, which is $I_0 = \langle S \rangle = \langle (\mathrm{Re}\, E_x)^2 \rangle_t/4\pi = E_0^2 \langle \cos^2 \omega t \rangle_t/4\pi = E_0^2/8\pi$. Similarly, on the basis of (5.17), the scattered intensity will be

$$I = \frac{1}{4\pi}\langle(\mathrm{Re}\,E)^2\rangle_t = \left(\frac{r_0}{r}\right)^2 \sin^2 \alpha \frac{E_0}{8\pi} = I_0 \left(\frac{r_0}{r}\right)^2 \sin^2 \alpha \tag{5.18}$$

The ratio between the power scattered by the electron on the surface element at distance r, $\mathrm{d}S = r^2\mathrm{d}\Omega$ (Ω is the solid angle) and the average incident power per unit surface is defined as the *cross-section* (Fig. 5.3):

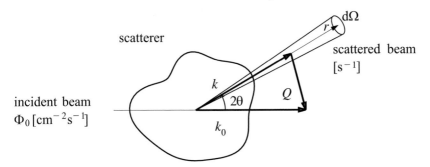

Fig. 5.3. Portion of the solid angle $\mathrm{d}\Omega$ seen by the detector in $\mathbf{r}$, for a scattering angle 2θ and a corresponding scattering vector $\mathbf{Q}$

$$\mathrm{d}\sigma \equiv \frac{Ir^2\mathrm{d}\Omega}{I_0} \tag{5.19}$$

From a corpuscular point of view, because the energy is transported by photons, $\mathrm{d}\sigma$ can be also viewed as the ratio between the number of photons per unit time that fall on the surface $\mathrm{d}S$ and the so-called photon-flux Φ_0, i.e. the number of incident photons per unit time and per unit area, $I_0 = \hbar\omega\Phi_0$. Moreover, since the frequency of the scattered beam is equal to that of the incident beam, we can say that the interaction between the photon and the electron has not changed the photon energy: we are dealing with an elastic collision. Combining (5.19) with (5.18), the *differential cross-section* is also defined:

$$\frac{\mathrm{d}\sigma}{\mathrm{d}\Omega} = r_0^2 \sin^2 \alpha \tag{5.20}$$

It is easy to show that in the case of unpolarized X-rays, the *Thompson scattering* of an electron reads

$$\frac{\mathrm{d}\sigma}{\mathrm{d}\Omega} = r_0^2 \frac{1 + \cos^2(2\theta)}{2} \tag{5.21}$$

where 2θ is the *scattering angle*, i.e. the angle between the direction of the incident beam z and the vector $\mathbf{r}$. The square root of $\frac{d\sigma}{d\Omega}$ is called the scattering length of the electron

$$b_X = r_0 \left[\frac{1 + \cos^2(2\theta)}{2} \right]^{1/2} \approx r_0 \quad \text{for } \theta \text{ small} \tag{5.22}$$

The total *Thompson scattering* cross-section is the integral over $d\Omega$, which reads

$$\sigma = \int d\Omega \frac{d\sigma}{d\Omega} = r_0^2 \int_0^{2\pi} d\psi \int_0^{\pi} d\theta \sin(2\theta) \frac{1 + \cos^2(2\theta)}{2} = \frac{8\pi}{3} r_0^2 \tag{5.23}$$

It can be observed from (5.16) that the scattered field $E(\mathbf{r}, t)$ is inversely proportional to the electron mass m. All other particles, starting from the proton, have a mass bigger than $\approx 2000m$. Hence, their contribution to the elastic scattering of X-rays is negligible.

If the average position of the electron is now in the arbitrary position $\mathbf{r}_n$, it is straightforward to show that the Thompson scattering (5.17) transforms to

$$E_n(\mathbf{r}, t) = -E_0 \, e^{-\imath \omega t} \frac{e^{\imath k_0 |\mathbf{r} - \mathbf{r}_n|}}{|\mathbf{r} - \mathbf{r}_n|} e^{\imath \mathbf{k}_0 \cdot \mathbf{r}_n} r_0 \sin \alpha \tag{5.24}$$

and, for unpolarized X-rays,

$$E_n(\mathbf{r}, t) = -E_0 \, e^{-\imath \omega t} \frac{e^{\imath k_0 |\mathbf{r} - \mathbf{r}_n|}}{|\mathbf{r} - \mathbf{r}_n|} e^{\imath \mathbf{k}_0 \cdot \mathbf{r}_n} b_X \tag{5.25}$$

Notice that in these equations $\mathbf{r}$ still represents the position of, say, the detector with respect to the origin of the reference frame. Let us suppose now that the distance r_n is orders of magnitude lower than the distance r. The first is in fact of the order of some Å's, the latter of some metres. We can thus write $|\mathbf{r} - \mathbf{r}_n| \approx r - \hat{\mathbf{r}} \cdot \mathbf{r}_n$, with $\hat{\mathbf{r}}$ the versor of $\mathbf{r}$. The spherical wave in (5.25) becomes

$$\frac{e^{\imath k_0 |\mathbf{r} - \mathbf{r}_n|}}{|\mathbf{r} - \mathbf{r}_n|} \approx \frac{e^{\imath k_0 r}}{r} e^{-\imath \mathbf{k} \cdot \mathbf{r}_n} \tag{5.26}$$

where, as shown in Fig. 5.4, $\mathbf{k} \equiv k_0 \hat{\mathbf{r}}$. The amplitude of the electric field scattered by the electron in $\mathbf{r}_n$ becomes

$$E_n(\mathbf{r}, t) \approx -E_0 \, e^{-\imath \omega t} \frac{e^{\imath k_0 r}}{r} b_X \, e^{\imath \mathbf{Q} \cdot \mathbf{r}_n} \equiv E(\mathbf{r}, t) e^{\imath \mathbf{Q} \cdot \mathbf{r}_n} \tag{5.27}$$

where we have introduced the so-called *scattering vector* $\mathbf{Q} \equiv \mathbf{k}_0 - \mathbf{k}$, with modulus $Q = 4\pi \sin \theta / \lambda$ (Fig. 5.4). The factor $e^{\imath \mathbf{Q} \cdot \mathbf{r}_n}$, called the *structure factor* of the electron in $\mathbf{r}_n$, represents the ratio between the amplitudes of two electromagnetic fields, one generated by the electron in $\mathbf{r}_n$ and one of the electron at the origin.

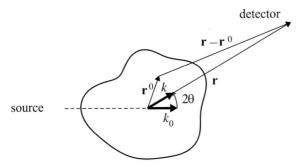

Fig. 5.4. Scattering geometry for the calculation of scattering at the detector in **r**

Instead of an electron, we now study the interaction between X-rays and a "blob" of matter, which will be viewed by X-rays as a set of electrons. We suppose that the scattering originated by each electron is weak, and only due to the interaction with the direct beam and not to the X-ray waves generated by the other electrons. This approximation is called the *single scattering approximation*. In **r** the amplitude of **E** will be due to the sum among all the amplitudes of single electrons. The structure factor will be

$$F(\mathbf{Q}) = \sum_{n=1}^{N} e^{\imath \mathbf{Q} \cdot \mathbf{r}_n} \tag{5.28}$$

According to (5.22), the scattering length can be defined as

$$A(\mathbf{Q}) = b_X F(\mathbf{Q}) = b_X \sum_{n=1}^{N} e^{\imath \mathbf{Q} \cdot \mathbf{r}_n} \tag{5.29}$$

and, similarly to (5.21), the cross-section is

$$\frac{d\sigma}{d\Omega}(\mathbf{Q}) = |A(\mathbf{Q})|^2 \equiv b_X^2 \sum_{m,n=1}^{N} e^{\imath \mathbf{Q} \cdot \mathbf{r}_{mn}} \tag{5.30}$$

where $\mathbf{r}_{mn} \equiv \mathbf{r}_m - \mathbf{r}_n$.

More realistically, the position of electrons is not exactly known; introducing the electron density function $\rho_e(\mathbf{r})$, we can write

$$F(\mathbf{Q}) = \int d^3\mathbf{r}\, \rho_e(\mathbf{r})\, e^{\imath \mathbf{Q} \cdot \mathbf{r}} \tag{5.31}$$

From a mathematical point of view, the latter integral is the so-called *Fourier transform* of $\rho_e(\mathbf{r})$. There is also the inverse Fourier transform, that allows the determination of $\rho_e(\mathbf{r})$ when $F(\mathbf{Q})$ is known:

$$\rho_e(\mathbf{r}) = \frac{1}{(2\pi)^3} \int d^3\mathbf{Q}\, F(\mathbf{Q})\, e^{-\imath \mathbf{Q} \cdot \mathbf{r}} \tag{5.32}$$

We can also add an equation which gives the scattering length of a blob of matter in an integral form:

$$A(\mathbf{Q}) = b_X \int d^3\mathbf{r}\, \rho_e(\mathbf{r})\, e^{i\mathbf{Q}\cdot\mathbf{r}} \tag{5.33}$$

Let us assume now that our blob of matter is an atom. If this atom is at the centre of the reference frame, with good approximation, its electronic density will have a spherical symmetry, $\rho_e(\mathbf{r}) \equiv \rho_e(r)$, where the polar coordinates of $\mathbf{r}$ are (r, θ, ϕ). Atomic structure factors become real functions that only depend on the modulus Q:

$$
\begin{aligned}
f(\mathbf{Q}) &= \int\limits_0^\infty r^2 dr \int\limits_0^\pi \sin\theta d\theta \int\limits_0^{2\pi} d\phi\, \rho_e(r)\, e^{i\mathbf{Q}\cdot\mathbf{r}} \\
&= 4\pi \int\limits_0^\infty r^2 dr \rho_e(r)\, \frac{\sin(Qr)}{Qr} \equiv f(Q)
\end{aligned}
\tag{5.34}
$$

In Fig. 5.5 calculated structure factors of some atoms are reported. It can be easily shown that if now the atom is translated in the position $\mathbf{r}_n$, its structure factor transforms to $f(Q)e^{i\mathbf{Q}\cdot\mathbf{r}_n}$. Hence, if we now take as blob of matter a

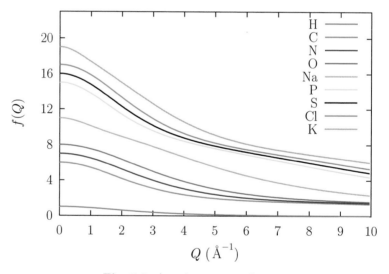

Fig. 5.5. Atomic structure factors

molecule and, as a first approximation, we suppose that the electronic clouds around the atoms do not change, the molecular form factor and the molecular scattering length (5.31) and (5.33) reduce to

$$F(\mathbf{Q}) = \sum_{n=1}^N f_n(Q)\, e^{i\mathbf{Q}\cdot\mathbf{r}_n} \tag{5.35}$$

$$A(\mathbf{Q}) = b_X \sum_{n=1}^{N} f_n(Q) \, e^{i\mathbf{Q}\cdot\mathbf{r}_n} \tag{5.36}$$

Neutrons

We turn now to study the interaction between neutrons and matter. In particular, we only describe the elastic scattering of neutrons. The energy of incident and scattered neutrons remains unchanged and its value is expressed by (5.9). In this condition, neutrons could only transfer momentum to the material system. The time-independent Schrödinger equation (5.5) can be written as

$$(\nabla^2 + k_0^2)\psi(\mathbf{r}) = \frac{2m_N}{\hbar^2} \, U(\mathbf{r})\,\psi(\mathbf{r}) \tag{5.37}$$

The latter is a second-order nonhomogeneous differential equation. It can be transformed to an integral equation with the method of Green's functions:

$$\psi(\mathbf{r}) = \psi^0(\mathbf{r}) + \frac{2m_N}{\hbar^2} \int d\mathbf{r}' G(\mathbf{r}, \mathbf{r}')\,\psi(\mathbf{r}')\,U(\mathbf{r}') \tag{5.38}$$

where $\psi^0(\mathbf{r})$ stands for the unperturbed state (5.7) and $G(\mathbf{r}, \mathbf{r}')$ represents a spherical wave with the same k_0, originating from the position $\mathbf{r}'$,

$$G(\mathbf{r}, \mathbf{r}') = \frac{e^{ik_0|\mathbf{r}-\mathbf{r}'|}}{4\pi|\mathbf{r}-\mathbf{r}'|} \tag{5.39}$$

The sense of (5.38) is clear: after the elastic collision, the new state $\psi(\mathbf{r})$ results from a superposition of the unperturbed wave $\psi^0(\mathbf{r})$ and spherical waves that are generated in $\mathbf{r}'$ with an amplitude depending on the value of $\psi(\mathbf{r}')$ and the potential $U(\mathbf{r}')$. Some approximations can be introduced. First, the range of the distances where the potential $U(\mathbf{r}')$ is different from zero will be of the typical order of atomic distances, say tenths of a nanometre. Hence, since we wish to calculate the unknown function $\psi(\mathbf{r})$ at the typical sample-to-detector distance r, say metres, the inequality $r' \ll r$ will be true. On the basis of (5.26) we have

$$G(\mathbf{r}, \mathbf{r}') \approx \frac{e^{ik_0 r}}{4\pi r} e^{-i\mathbf{k}\cdot\mathbf{r}'} \tag{5.40}$$

Second, (5.38) is only a formal solution of the Schrödinger equation, with $\psi(\mathbf{r})$ in both members still unknown. Nevertheless, an iterative procedure can be written,

$$\psi = \psi^0 + \widehat{G}U\psi \approx \psi^0 + \widehat{G}U\psi^0 + \widehat{G}U\widehat{G}U\psi^0 + \dots$$
$$= \psi^0 + \psi^1 + \dots \tag{5.41}$$

where $\widehat{G}$, defined as Green's operator, represents the integral in (5.38). Each step can be thought of as an elastic scattering event. In the so-called *Born approximation* only the first term $\psi^1 = \widehat{G} U \psi^0$ is considered. Similarly to the X-rays, this view is also called "single scattering approximation" and the smaller the thickness of the sample that neutrons cross, the more valid the Born approximation. Combining (5.38), (5.40) and (5.7), the first-order approximation reads

$$\psi^1(\mathbf{r}) = \frac{1}{L^{3/2}} \frac{e^{ik_0 r}}{r} \frac{1}{4\pi} \frac{2m_N}{\hbar^2} \int d\mathbf{r}' \, U(\mathbf{r}') \, e^{i\mathbf{Q}\cdot\mathbf{r}'} \tag{5.42}$$

where the scattering vector $\mathbf{Q} \equiv \mathbf{k}_0 - \mathbf{k}$ represents the momentum (in $\hbar$ units) that the neutron has transferred to the sample (Fig. 5.4). For the sake of completeness, we also report the first-order approximation of the whole wave function,

$$\Psi^1(\mathbf{r}, t) = \frac{1}{L^{3/2}} e^{-i\frac{\mathcal{E}}{\hbar}t} \frac{e^{ik_0 r}}{r} \frac{1}{4\pi} \frac{2m_N}{\hbar^2} \int d\mathbf{r}' \, U(\mathbf{r}') \, e^{i\mathbf{Q}\cdot\mathbf{r}'} \tag{5.43}$$

Neutrons mainly interact with nuclei. The interaction potential due to a nucleus at position $\mathbf{r}_n$ is described by *Fermi's pseudo-potential*

$$U(\mathbf{r}) = \frac{2\pi\hbar^2}{m_N} b_n \delta(\mathbf{r} - \mathbf{r}_n) \tag{5.44}$$

where b_n is the *scattering length* of the nth nucleus and $\delta(\mathbf{r})$ is the well-known three-dimensional Dirac delta function. The Fourier transform in (5.43) is straightforward:

$$\Psi_n^1(\mathbf{r}, t) = \frac{1}{L^{3/2}} e^{-i\frac{\mathcal{E}}{\hbar}t} \frac{e^{ik_0 r}}{r} b_n \, e^{i\mathbf{Q}\cdot\mathbf{r}_n} \tag{5.45}$$

Note the similarity with Eq. (5.27), which, for X-rays, gives the amplitude of the electric field scattered by the electron in $\mathbf{r}_n$. This analogy makes possible a unique mathematical treatment of the two types of radiation.

The incident neutron-flux over a box of side L will be $\Phi_0 = v_N / L^3$. Considering the meaning in quantum-mechanics of Ψ, the cross-section due to the neutron–nucleus elastic interaction, similarly to (5.19), is

$$d\sigma \equiv \frac{v_N |\Psi_n^1(\mathbf{r}, t)|^2 r^2 d\Omega}{\Phi_0} = b_n^2 d\Omega \tag{5.46}$$

Hence, the differential and the total scattering cross-section are

$$\frac{d\sigma}{d\Omega} = b_n^2 \tag{5.47}$$

$$\sigma = 4\pi b_n^2 \tag{5.48}$$

Turning now to a blob of matter, say a molecule, its interaction with a neutron beam will be described by the sum of all the Fermi potentials due to each nth atom. Hence, (5.45) becomes

$$\Psi_n^1(\mathbf{r}, t) = \frac{1}{L^{3/2}} e^{-i\frac{\xi}{\hbar}t} \frac{e^{ik_0 r}}{r} \sum_{n=1}^{N} b_n e^{i\mathbf{Q}\cdot\mathbf{r}_n} \tag{5.49}$$

and the scattering length of the system will be

$$A(\mathbf{Q}) = \sum_{n=1}^{N} b_n e^{i\mathbf{Q}\cdot\mathbf{r}_n} \tag{5.50}$$

or, turning to an integral form, we find a deep analogy with (5.33)

$$A(\mathbf{Q}) = \int d^3\mathbf{r}\, \rho(\mathbf{r})\, e^{i\mathbf{Q}\cdot\mathbf{r}} \tag{5.51}$$

where $\rho(\mathbf{r})$ is defined as the *scattering length density*

$$\rho(\mathbf{r}) = \lim_{V\to 0} \frac{1}{V} \sum_{n=\text{atoms in } V}^{N} b_n = \sum_{n=1}^{N} b_n \delta(\mathbf{r} - \mathbf{r}_n) \tag{5.52}$$

5.2 Microstructural Investigations by Small Angle Scattering of Neutrons and X-rays

5.2.1 Introduction

Small Angle Scattering techniques [74, 88, 90] (SANS when using neutron beams or SAXS when using conventional X-ray radiation sources or synchrotron radiation) are experimental methods allowing the determination of structural features, such as size and volume fraction, of matrix inhomogeneities in a huge variety of materials, covering studies from biochemistry and biophysics to applied and industrial research. The order of magnitude of the inhomogeneities size that can be detected is in the approximate range 1–10^3 nm, but with special methods (Ultra-SANS) also inhomogeineities up to tens of μm can be investigated.

They are usually considered as complementary techniques to *Transmission Electron Microscopy* (TEM). In fact, correct use of SANS/SAXS is often based on some information that can be obtained from TEM, for instance, the particle shape. On the other hand, SANS/SAXS are performed on cross-sectional areas of some mm^2, or even cm^2, so that their results are much more significant, from a statistical point of view, than those obtained by TEM (few μm^2). Moreover they can be used also for kinetics investigations, owing to their nondestructive character.

5.2.2 Theoretical Bases

Cross-sections

In general it can be shown, by extrapolating to a "blob" of matter containing different atoms (5.30), which is valid for a "blob" containing electrons, that in scattering processes ruled by the interference of waves scattered by the material atoms, the differential cross-section is given by

$$\frac{d\sigma}{d\Omega} = \sum_i \sum_j b_i b_j \exp\left[\imath \mathbf{Q} \cdot (\mathbf{r}_i - \mathbf{r}_j)\right] \tag{5.53}$$

where the sum is extended to all atoms hit by the beam, each with spatial position $\mathbf{r}_i$, b_i being the scattering length of the ith atom. In the case of X-rays, b is related to the electronic properties of the target atom ($b = Zr_e$, where Z is the atomic number and r_e is the classical electron radius), while in the case of neutrons it depends on the nuclear interactions with the target isotope. Therefore, b varies regularly through the periodic table in X-ray scattering, but it does not in neutron scattering. Furthermore, in the latter case, for some isotopes b can also assume negative values.

$\mathbf{Q}$ is the scattering vector, given by the difference between the wave vector of the scattered ($\mathbf{k}$) and incident ($\mathbf{k}_0$, with $|\mathbf{k}_0| = |\mathbf{k}|$) radiation (Fig. 5.3). The cross-section and the intensity of the scattered radiation depend on the scattering angle 2θ, but usually these quantities are expressed as functions of the modulus of $\mathbf{Q}$, as the latter is linked to 2θ by the relation

$$Q = \frac{4\pi}{\lambda} \sin\theta \tag{5.54}$$

where λ is the wavelength of the incident radiation.

For low-Q values the single atoms are not resolved anymore, and the process is ruled by the interference of the waves scattered from regions with linear dimensions $R \approx \pi/Q$. In this case the sum in (5.53) becomes an integral (Fourier transform)

$$\frac{d\sigma}{d\Omega} = \left|\int_V \rho(\mathbf{r}) \exp(\imath \mathbf{Q} \cdot \mathbf{r}) \, d^3\mathbf{r}\right|^2 \tag{5.55}$$

where $\rho(\mathbf{r})$ is the local scattering length density. It is convenient to express $\rho(\mathbf{r})$ in terms of its fluctuations around the mean value

$$\rho(\mathbf{r}) = \langle\rho\rangle + \delta\rho(\mathbf{r}) \tag{5.56}$$

For $Q > 0$ the contribution of the constant term $\langle\rho\rangle$ vanishes, and (5.3) becomes

$$\frac{d\sigma}{d\Omega} = \left|\int_V \delta\rho(\mathbf{r}) \exp(\imath \mathbf{Q} \cdot \mathbf{r}) \, d^3\mathbf{r}\right|^2 \tag{5.57}$$

Two-phase model and form factor

In most cases it is possible to assume that the system is formed by particles (such as precipitates or cavities) with homogeneous scattering length density

ρ_p, embedded in a matrix with homogeneous scattering length density ρ_m. With this assumption, (5.57) becomes

$$\frac{d\sigma}{d\Omega} = (\rho_p - \rho_m)^2 \left| \int_{V_p} \exp\left(\imath \mathbf{Q} \cdot \mathbf{r}\right) d^3\mathbf{r} \right|^2 \tag{5.58}$$

where V_p is the particle volume and $(\rho_p - \rho_m)^2 = (\Delta\rho)^2$ is the nuclear contrast between the particle and matrix. If the N_p particles are identical, using the macroscopic cross-section $d\Sigma/d\Omega = (N_p/V)(d\sigma/d\Omega)$, where V is the sample volume, (5.58) can be written as

$$\frac{d\Sigma}{d\Omega} = V_p^2 \left(\frac{N_p}{V}\right) (\Delta\rho)^2 \left|F_p\left(\mathbf{Q}\right)\right|^2 \tag{5.59}$$

where $F_p\left(\mathbf{Q}\right)$ is the form factor, defined as

$$F_p\left(\mathbf{Q}\right) = \frac{1}{V_p} \left| \int_{V_p} \exp\left(\imath \mathbf{Q} \cdot \mathbf{r}\right) d^3\mathbf{r} \right| \tag{5.60}$$

In general $\left|F_p\left(\mathbf{Q}\right)\right|^2$ depends on, besides the particle geometry, the direction of $\mathbf{Q}$ with respect to the particle principal axes. Anyway, in the case of randomly oriented particles, the observed form factor is an average over the whole solid angle, and $\left|F_p\left(\mathbf{Q}\right)\right|^2$ depends only on the modulus of $\mathbf{Q}$. For spherical particles with radius R (see also Fig. 5.6)

$$F_p\left(\mathbf{Q}\right) = 3\frac{\sin\left(QR\right) - QR\cos\left(QR\right)}{\left(QR\right)^3} \tag{5.61}$$

Polydispersion

Let us consider the case in which the particles are not identical, but are polydispersed with some size distribution $N\left(R\right)$, defined as the number of particles in unit volume with size between R and $R + dR$. In that case (5.59) becomes

$$\frac{d\Sigma}{d\Omega} = (\Delta\rho)^2 \int_0^\infty N\left(R\right) V^2\left(R\right) \left|F\left(QR\right)\right|^2 dR \tag{5.62}$$

where the "p" index has been omitted, as all the quantities under the integral are referred to the particles.

With the definition given above for $N\left(R\right)$, the particle concentration (c) and volume fraction (f) can be calculated as

$$c = \int_0^\infty N\left(R\right) dR \qquad f = \int_0^\infty N\left(R\right) V\left(R\right) dR \tag{5.63}$$

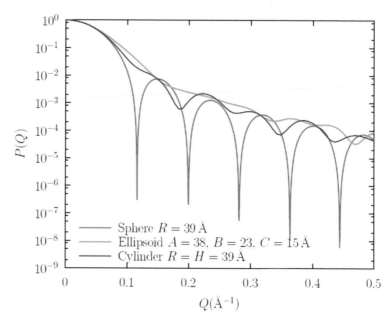

Fig. 5.6. Examples of form factors for different particle geometries

Coherent and incoherent neutron scattering

In the case of neutron beams, the interaction can take place with all of the isotopes corresponding to the same chemical element. The neutron–isotope interaction will depend on the total angular momentum $J = I + s$, where I is the operator associated with the isotope spin $I \neq 0$ and s is the operator associated with the neutron spin $1/2$. Hence, there are two different scattering lengths, $b_{\pm}$, associated with the two different values $J_{\pm} = I \pm 1/2$. On this basis, (5.53) becomes

$$\frac{\mathrm{d}\sigma}{\mathrm{d}\Omega} = \sum_i \sum_j \langle b_i b_j \rangle \, \langle \exp\left[\imath \mathbf{Q} \cdot (\mathbf{r}_i - \mathbf{r}_j)\right]\rangle \tag{5.64}$$

which can be transformed into

$$\frac{\mathrm{d}\sigma}{\mathrm{d}\Omega} = \sum_i \sum_j b_i^{\mathrm{coh}} b_j^{\mathrm{coh}} \, \langle \exp\left[\imath \mathbf{Q} \cdot (\mathbf{r}_i - \mathbf{r}_j)\right]\rangle + \sum_i \left(b_i^{\mathrm{incoh}}\right)^2 \tag{5.65}$$

where

$$\begin{aligned} b_i^{\mathrm{coh}} &= \frac{1}{2I+1}\left[(I+1)\,b_+ + I b_-\right] \\ b_i^{\mathrm{incoh}} &= \frac{I(I+1)}{2I+1}\,(b_+ - b_-)^2 \end{aligned} \tag{5.66}$$

Therefore, the cross-section for small angle neutron scattering consists of two terms: the first one, depending on Q, contains the structural information about the particles in the sample matrix, while the second one, not depending on Q, is the incoherent background of the sample.

Low-Q and high-Q behaviours

In the case of $QR < 1.5$ (about), where R is some *characteristic length* of the particles, it can be shown that, independently on the particle shape, the expansion of the form factor in Taylor's series leads to the following relation (*Guinier's approximation*):

$$\frac{\mathrm{d}\Sigma}{\mathrm{d}\Omega} = V_p^2 \left(\frac{N_p}{V}\right) (\Delta\rho)^2 \exp\left(-\frac{Q^2 R_g^2}{3}\right) \tag{5.67}$$

where

$$R_g^2 = \frac{1}{V_p} \int_{V_p} r^2 \mathrm{d}^3\mathbf{r} \tag{5.68}$$

is the gyration radius (squared) of the particle and r is the distance of the volume element $\mathrm{d}^3\mathbf{r}$ from the centre of mass. R_g depends on the particle geometry; as an example, for a sphere of radius R, (5.68) gives $R_g = \sqrt{3/5}R$. In the case of polydispersion it can be shown that Guinier's approximation is still valid, but the gyration radius depends on the size distribution. For spheres with radius R

$$R_g^2 = \frac{3\langle R^8 \rangle}{5\langle R^6 \rangle} = \frac{3\int\limits_0^\infty N(R) R^8 \mathrm{d}R}{5\int\limits_0^\infty N(R) R^6 \mathrm{d}R} \tag{5.69}$$

According to (5.15), by plotting $\ln(\mathrm{d}\Sigma/\mathrm{d}\Omega)$ versus Q^2 ("Guinier's plot") one obtains a straight line, with slope $R_g^2/3$ and intercept $(N_p/V) V_p^2 (\Delta\rho)^2$. At high Q values (that is, for $QR > 3$), the "Porod's approximation" holds for sharp particle–matrix interfaces

$$\frac{\mathrm{d}\Sigma}{\mathrm{d}\Omega} = \frac{2\pi (\Delta\rho)^2 S_p}{Q^4} \tag{5.70}$$

where S_p is the particle specific surface.

Kratky plot and Porod's invariant

A well-used method to represent the small angle scattering data is the so-called "*Kratky plot*", that is $Q^2 (\mathrm{d}\Sigma/\mathrm{d}\Omega)$ versus Q. This kind of plot is useful for a number of reasons, among which we mention: (i) if Guinier's approximation is valid, the Kratky plot has a maximum at $Q = \sqrt{3}/R_g$, so that the gyration radius can be directly determined by the peak position; (ii) from the integral of the Kratky plot (the so-called "Porod's invariant" $\widetilde{Q}$), the volume fraction f of the particles can be calculated according to the relation

$$\widetilde{Q} = \int_0^\infty \frac{\mathrm{d}\Sigma}{\mathrm{d}\Omega}(Q)\, Q^2 \mathrm{d}Q = 2\pi^2\, (\Delta\rho)^2\, f\, (1-f) \qquad (5.71)$$

Nondiluted systems: the structure factor

When the system of particles inside the matrix is highly concentrated, an interference between waves scattered from neighbour particles takes place. This phenomenon is taken into account by a further factor $S(Q)$ ("structure factor" $S(Q) = 1$ for diluted systems) appearing in (5.59) (or (5.62) in the case of polydispersion):

$$\frac{\mathrm{d}\Sigma}{\mathrm{d}\Omega} = V_p^2 \left(\frac{N_p}{V}\right) (\Delta\rho)^2\, |F_p\,(\mathbf{Q})|^2\, S\,(Q) \qquad (5.72)$$

$$\frac{\mathrm{d}\Sigma}{\mathrm{d}\Omega} = (\Delta\rho)^2 \left[\int_0^\infty N\,(R)\, V^2\,(R)\, |F\,(QR)|^2\, \mathrm{d}R\right] S\,(Q) \qquad (5.73)$$

$S(Q)$ is linked to the Fourier transform of the "correlation function" $g(r)$, which gives the probability of finding two particles with centres at a distance r from each other. For spherical particles (Fig. 5.7)

$$S\,(Q) = 1 + \frac{N_p}{V} \int_0^\infty 4\pi r^2\, [g\,(r) - 1]\, \frac{\sin(Qr)}{Qr}\mathrm{d}r \qquad (5.74)$$

The combination of form factor and structure factor, in the case of polydispersion, leads to the appearance of a peak in the plot $(\mathrm{d}\Sigma/\mathrm{d}\Omega)$ versus Q (Fig. 5.7), the position of which is inversely proportional to the mutual distance of the centres of first neighbour particles.

5.2.3 Experimental Methods and Data Analysis

Experimental set-up

The standard set-up for a SANS experiment is shown in Fig. 5.8. Its main principles are the same also in the case of SAXS. The incident radiation, monochromatized by a mechanical velocity selector (for neutrons) or by Bragg diffraction from a perfect crystal (both for neutrons and X-rays), is collimated and sent onto the sample, where the scattering from matrix inhomogeneities takes place. The scattered beam is collected in a 2D position sensitive detector. In the case of neutrons, this is formed by a matrix of gas detectors (usually

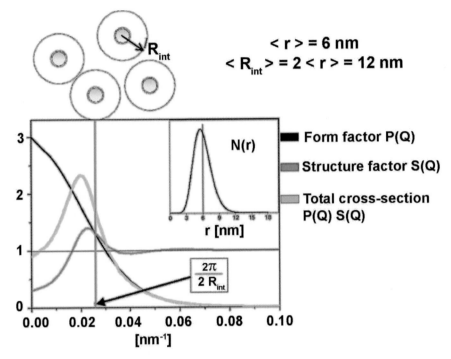

Fig. 5.7. Effect of polydispersion and structure factor on the total cross-section (spherical particles)

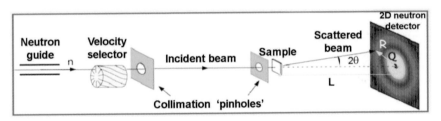

Fig. 5.8. Experimental set-up of a SANS experiment

64×64 cells, 1 cm^2 each, or 128×128 cells, 0.25 cm^2 each); in the case of X-rays the 2D detector can be a system formed by an X-ray image intensifier and a CCD camera [132]. Referring to Fig. 5.8, the Q value corresponding to each cell of the 2D detector is given by

$$Q = \frac{4\pi}{\lambda} \sin\theta \approx \frac{2\pi}{\lambda} \frac{R}{L} \qquad (5.75)$$

where R is the radial position of the cell with respect to the incident beam axis, and L is the sample–detector distance. Therefore, the Q-range can be chosen acting on either λ or L (or both). The maximum Q is determined by the 2D detector size, while the minimum is determined by the size of the beam stopper, that is an absorber put in front of the detector, in the region of the incident beam axis, to prevent damage by direct radiation reaching the detector with no interaction inside the sample.

As the maximum detectable particle size is roughly given by $R_{\max} \approx 2\pi/Q_{\min}$, it is evident that the presence of a beam stopper is a limit to the possibility to investigate "big" particles, such as, for instance, voids and cavities (tens of μm). This can be overcome using a *double-crystal diffractometer* (Fig. 5.9), to obtain the so-called Ultra-SANS (USANS) [114]. The neutron beam, monochromatized by the Bragg diffraction at monochromator, enters the end face of an asymmetrically cut analyser crystal and propagates along its longest edge. The analyser crystal reflects the neutron beam only in the case the scattered beam fulfils the Bragg condition. For neutrons scattered by a sample this takes place at different parts of the bent analyser crystal, so that the scattering diagram can be recorded as a whole by the position-sensitive detector. It is possible to change curvatures of the monochromator and the analyser within a rather wide range. In this way one can modify both the resolution (monochromator) and the Q-range available (analyser).

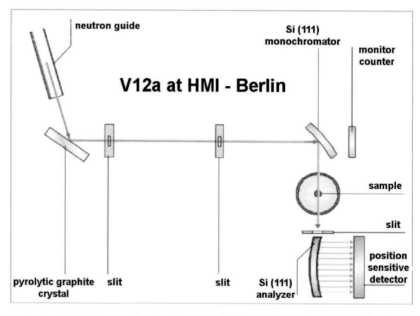

Fig. 5.9. Scheme of a double-crystal diffractometer for USANS [114]

Data analysis and the inverse problem

As stated in previous sections, the behaviours at low Q (Guinier's approximation) and high Q (Porod's approximation) can give information about the gyration radius and the specific surface of the particles. The gyration radius can also be determined by the peak position in the Kratky plot, the integral of which can be used to determine the particle volume fraction (5.71).

In the case of polydispersion, in the framework of the two-phase model, (5.62) can be fitted to the experimental data to determine the size distribution $N(R)$, provided the particle shape, and thus the form factor, is known (e.g. from electron microscopy observations). In this case some shape for $N(R)$ should be assumed. According to the system under investigation, Gaussian, log-normal, Weibull, Pearson or other distributions can be used. If no particular distribution shape can be assumed *a priori*, or also when more than one particle family could be present, then a common method is to write $N(R)$ as a linear combination of N_S β-spline functions

$$N(R) = \sum_{i=1}^{N_S} c_i \beta_i(R) \tag{5.76}$$

so that

$$\frac{\mathrm{d}\Sigma}{\mathrm{d}\Omega} = (\Delta\rho)^2 \sum_{i=1}^{N_S} c_i \int_0^\infty \beta_i(R) V^2(R) |F(QR)|^2 \, \mathrm{d}R \tag{5.77}$$

The least square-fitting of (5.77) to the experimental data determines the c_i coefficients.

5.2.4 Recent Applications to Materials of Technological and Industrial Interest

Study of precipitation in 7XXX Al alloy

Many studies have recently focused on Li-containing 7XXX series (Al-Zn-Mg-Cu) alloys, for application in aerospace technology. The precipitation process in these alloys during various ageing conditions is of great importance for the understanding of their mechanical strength.

Du *et al.* [60] recently performed a SAXS experiment aiming to determine the size distribution of precipitates (η and η') induced by different thermal treatments. The following states were investigated: 24 h at 160°C (A), 48 h at 160°C (B), 72 h at 160°C (C), 96 h at 160°C (D), 24 h at 130°C (E), 24 h at 150°C (F), 3.5 h at 120°C + 15 h at 160°C (G). It was found that the best results are obtained assuming an average ellipsoidal shape of the precipitates, with an axis ratio between 0.2 and 0.4. Figure 5.10 shows the distribution (log-normal) obtained for the longest half-length of the ellipsoids, from which it was concluded that the average length and standard deviation increase with the "kinetic strength" K_s, that is a scaling variable depending on the time evolution of the temperature [91].

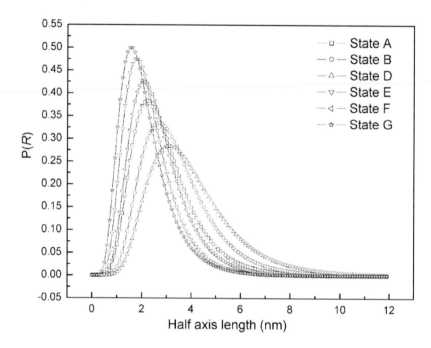

Fig. 5.10. Log-normal distributions of longest half-length for Li-containing Al-Zn-Mg-Cu alloy aged at different conditions [60]

Nanocrystalline MgH₂ for reversible storage of hydrogen

Nanocrystalline magnesium hydride is considered to be one of the most promising alternatives for the reversible storage of hydrogen. Pranzas et al. [182] recently investigated the structural changes of high-energy ball-milled MgH_x and MgH_2/Cr_2O_3 with varying hydrogen content, by small and ultra small-angle neutron scattering (SANS/USANS), using different milling parameters to obtain information about hydrogen sorption and desorption mechanisms. Results are shown in Fig. 5.11. The volume fraction distributions obtained indicate crystallite coarsening during cycling at 300°C, as well as breakup of particles of radii larger than 10 μm during the cycling process.

SiO₂ nanopowders

Nanopowders have found use in various applications, including fillers, catalysts and medicines. Amorphous SiO_2 particles, for instance, are used as the filler for a semiconductor chip-protective sealing agent that is called an epoxy moulding compound (EMC). When a filler is mixed with an epoxy resin its particle size distribution is critical, and therefore the need to establish a technique for particle size distribution analysis of nanopowders has become urgent.

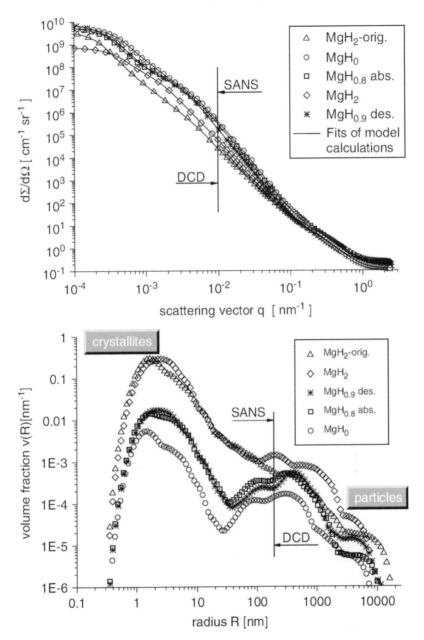

Fig. 5.11. Combined SANS/USANS curves (a) and volume fraction distributions for MgH$_x$ samples investigated in [182] (b)

Hashimoto et al. [96] used the USAXS technique to determine the size distribution in samples with different specific surface areas (SSA): 50, 130, 200 and 380 m^2/g were considered. The results show good agreement with TEM observations (Fig. 5.12).

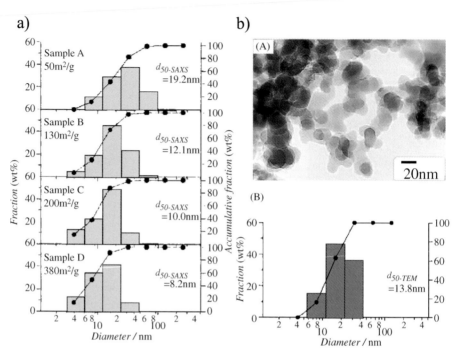

Fig. 5.12. Results for SiO$_2$ nanoparticles in the different samples investigated in [96]. (a) Size distribution obtained by USAXS; (b) TEM micrograph (A) and corresponding size distribution (B) for sample B

Creep cavitation damage in a stainless steel pressure vessel

Grain-boundary cavitation is the dominant failure mode associated with initiation of reheat cracking, which has been widely observed in austenitic stainless steel pressure vessels operating at temperatures within the creep range ($> 450°C$). SANS and USANS experiments at the LLB PAXE instrument (Saclay) and the V12 double-crystal diffractometer of the HMI-BENSC facility (Berlin), respectively, were used to characterize cavitation damage in a variety of creep specimens extracted from ex-service nuclear power plant [35]. The results for the SANS experiment are shown in Fig. 5.13. The SANS curves (Fig. 5.13a) were treated as a superimposition of a contribution from the Porod scattering (given by the cavities) and a signal at higher Q from smaller particles (precipitates). The specific surface of the cavities was then evaluated, which is proportional to the fractional grain boundary creep damage (Fig. 5.13b). The cavity size distributions and volume fractions where ob-

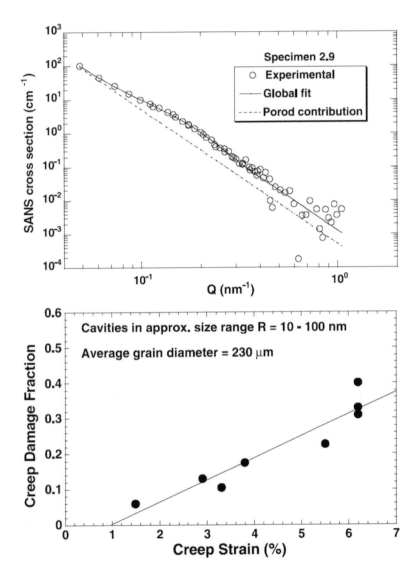

Fig. 5.13. SANS cross section (a) and fractional grain boundary creep damage obtained (b) [24]

tained from the USANS experiment (Fig. 5.14a), which also confirmed the approximately linear increase of the damage with creep strain level (Fig. 5.14b).

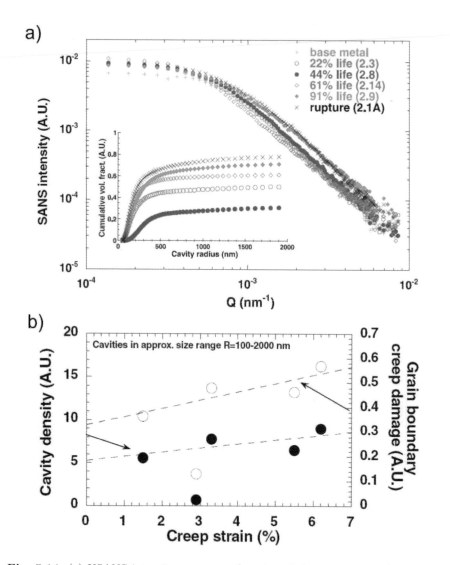

Fig. 5.14. (a) USANS intensity curves, as function of the creep strain (inset: the cumulative volume fractions obtained); (b) cavity density and grain boundary creep damage [182]

Precipitation in NiTi shape-memory alloys

The *martensitic phase transitions* (MT) in ordered body centred cubic (B2-structured) NiTi shape memory alloys depend strongly on composition and heat treatment. In the two-phase regime below the stability range, Ni-rich off-stoichiometric NiTi follows a decomposition sequence implying the presence of Ni4Ti3 precipitates, with a rhombohedral unit cell. These precipitates are known to improve shape memory behaviour, influencing the MT sequence and transformation temperatures [159, 194]. It is also known that for quenched samples (out of the NiTi phase stability range), a MT from the NiTi phase to the B19' phase (monoclinic structure) occurs up to about Ni - 49 at. % Ti. The microstructure in the "as quenched" state of five alloys, with different Ti content (nominal concentrations between 46 and 49 at. % Ti) was investigated by SANS [131]. Two annealed samples (1 h at 553 K after previous quenching in water from 1273 K) were also studied to provide a reference for a well-established microstructure, with Ni_4Ti_3 precipitates as revealed by transmission electron microscopy. Figure 5.15 shows the measured coherent

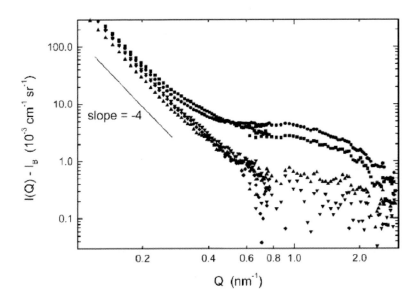

Fig. 5.15. Coherent SANS cross-sections for NiTi alloys with different Ti content: ($\Diamond$) Ni49 at. % Ti, ($\triangledown$) Ni48.4 at. % Ti, ($\triangle$) Ni47.9 at. % Ti, ($\square$) Ni47 at. % Ti, ($\bigcirc$) Ni46 at. % Ti [131]

SANS cross-sections for the five alloys studied. For the sample with 49 at. % Ti no significant additional scattering contribution is observed. For the other solution-treated samples a small increase of coherent scattering is observable.

Figure 5.16 shows the volume of a single precipitate V_p and the volume fraction C_p of these precipitates as a function of the average Ti concentration of the samples. Both V_p and C_p start to increase from about 49 at. % Ti with decreasing Ti concentration. As the MT to the B19' phase abruptly vanishes at about 49 at. % Ti for all lower Ti concentrations, associated with the detection of the Ni_4Ti_3 precipitates, it can be concluded that these precipitates, in an early stage of formation, suppress the MT for quenched Ni-rich alloys.

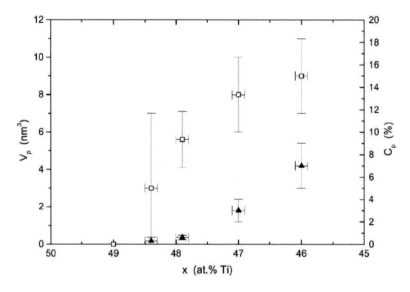

Fig. 5.16. ($\square$) Volume V_p of a single precipitate and ($\triangle$) volume fraction C_p as a function of the average Ti concentration of NiTi alloys after quenching from 1273 K in ice water [131]

5.3 Residual Stress Determination by Neutron and X-ray Diffraction

5.3.1 Introduction

Residual stresses are autobalancing stresses existing in a free body not submitted to any external surface force. These stresses arise from the elastic response of the material to a nonhomogeneous distribution of nonelastic strains such as plastic strains, precipitation, phase transformation, misfit, thermal expansion strains, etc. Several manufacturing processes and thermal or mechanical treatments leave residual stresses within the components. For example, mechanical processes that cause plastic deformation in the surface layers of the material, such as shot-peening, grinding, machining, etc., cause residual stresses in these layers because of the constraining effect of the bulk, where plastic deformation is minimal. These stresses are called *macrostresses*. Since the surface layers constrain the bulk in return, the bulk material is also subjected to residual stresses even if it may not have suffered deformation.

Residual stresses can be beneficial or detrimental, depending if they counteract or not external loads; for instance, when they add to external loads, they can accelerate the fatigue process and induce earlier failure of the component. Residual stresses and strains play a key role in several sectors of mechanical engineering, for instance:

- in welded structures (for ship construction, nuclear technology, aerospace technology, pipelines, oil rigs, etc.);
- in heat treatments of finished parts (ground gears, shot-peened, sand blasted pieces, material subjected to laser heat treatments or quenched after heat treatment, etc.);
- in the fatigue response of solids;
- in the phenomenon known as stress corrosion.

Let's consider first a material that undergoes no change in crystal structure during heat treatment. If aluminium is cooled quickly from the heat treatment temperature, the surface and the inner regions contract at different rates. That difference, coupled with low material yield strength associated with the high temperature, sometimes induces plastic flow or permanent yielding. In fact in the first step the surface regions, which tend to contract on cooling more than the bulk material because of the temperature gradient, are extended by the interior while, vice versa, the internal region is compressed, according to Newton's third law.

On the other hand on a second step on continued cooling to room temperature, the surface regions which have previously been extended end up in compression because the bulk shrinks. These residual surface stresses are important because every type of change in cross-section concentrates additional applied tensile stresses near the surface and can act to initiate a crack. Residual compressive stresses in the surface must be overcome by the applied load to initiate cracks. As a conclusion, in this case, the presence of surface compressive stresses is a highly favourable condition. Heat treatments do not always produce surface compressive stresses: an example is a material undergoing a phase transformation like steel hardened to martensite. In this process, the surface regions reach the transformation temperature, transform into martensite and expand while the interior ones, composed of low strength austenite, deform plastically to partially accommodate this change. After that, at a different temperature, the bulk material transforms into martensite producing an expansion which is hindered by the high strength martensitic surface. At a lower temperature, close to room temperature, the surface is put in tension by the bulk, producing surface residual tensile stresses, which contribute to crack initiation and propagation. One way to replace surface tensile stresses with compressive stresses is to peen the surface: an example will be considered below.

5.3.2 Residual Stress Classification

Usually three types of residual stresses are defined, i.e. first, second and third order, and are schematically shown in Fig. 5.17 for a polycrystalline sample.
Residual stresses of the first order, σ^{I} are averaged over several grains of the materials. These stresses are also often called *macrostresses*.
Residual stresses of the second order, σ^{II} are averaged over a single grain. In Fig. 5.17 the stresses of this type are shown for two adjacent grains.
Residual stresses of the third order, σ^{III} are inhomogeneous across a single grain. In Fig. 5.17 the stresses of this type are shown inside two grains.
In general the sums of the residual stresses of the second and third order are called *microstresses*. The total residual stress in a given point is given by the sum of the three types of stresses as shown in Fig. 5.17.

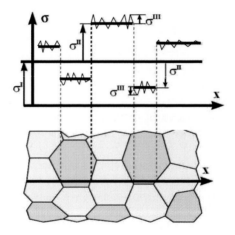

Fig. 5.17. Schematic representation of the three types of residual stresses

5.3.3 Techniques for Residual Stress Determination

The most widely used techniques to determine residual stresses are mentioned here:

- *hole drilling*: consisting in drilling a hole and measuring the subsequent relaxation around the hole (the distortion) with electrical-resistance strain gauges. It is a destructive technique;
- *acoustic wave propagation*: the wave velocity depends on the state of strain, ε, in the material, and hence on the residual stresses;

- *method associated with the Barkhausen effect*, which is based on the measurement of the perturbation of the magnetic field in ferromagnetic phases, induced by the presence of residual stresses;
- *methods based on X-ray or neutron diffraction*: they are the object of the present chapter and thus will be described in detail below.

5.3.4 Elastic Stress–Strain Relations

The method of residual stresses determination either by X-ray or neutron diffraction is actually based on the precise measurement of the strains. For a detailed and comprehensive treatment of these methods the reader should refer to [169] and [97]. Then by knowing the stress–strain relationship it is possible to deduce the residual stresses. In fact this relationship is Hooke's law, which has the general form

$$\varepsilon_{ij} = C_{ijkl}\sigma_{kl} \tag{5.78}$$

which can be inverted giving the stress as a function of the strain:

$$\sigma_{ij} = S_{ijkl}\varepsilon_{kl} \tag{5.79}$$

S_{ijkl} and C_{ijkl} are the stiffness and compliance moduli of the material and both are tensors of the fourth rank (1.6).

Each one of the Eqs. (5.78) and (5.79) represents in fact nine linear equations, each having nine terms, so that there are 81 constants in the S_{ijkl} and C_{ijkl} tensors. Both tensors satisfy the following group of symmetry $S_{ijkl} = S_{jikl} = S_{ijlk} = S_{klij}$ and $C_{ijkl} = C_{jikl} = C_{ijlk} = C_{klij}$ which reduces the number of constants to 21. But in fact the number of independent elastic constants is equal to 18 since 3 of them turn out to be directional cosines configuring eigendirections of material anisotropy in the global reference coordinate system. In the case of *orthotropy* the number of constants reduces to nine, and in the case of *transversely isotropic material*, to six (1.7).

However it can be shown that in the case of an *elastically isotropic material* the generalized Hooke's law (5.79) can be strongly simplified. In particular for a material with *Young modulus* E and *Poisson's ratio* ν, the principal internal stresses along the principal strain axes, $\sigma_{xx}, \sigma_{yy}, \sigma_{zz}$, are related to the strains by

$$\begin{aligned}
\sigma_{xx} &= \frac{E}{(1+\nu)(1-2\nu)}\left[(1-\nu)\,\varepsilon_{xx} + \nu\,(\varepsilon_{yy} + \varepsilon_{zz})\right] \\
\sigma_{yy} &= \frac{E}{(1+\nu)(1-2\nu)}\left[(1-\nu)\,\varepsilon_{yy} + \nu\,(\varepsilon_{xx} + \varepsilon_{zz})\right] \\
\sigma_{zz} &= \frac{E}{(1+\nu)(1-2\nu)}\left[(1-\nu)\,\varepsilon_{zz} + \nu\,(\varepsilon_{yy} + \varepsilon_{xx})\right]
\end{aligned} \tag{5.80}$$

From (5.80) it appears that in order to determine the stresses it is necessary to determine the strain tensor, once the elastic constants are known for the investigated material.

In fact in the most general case the strains and stresses are triaxal and can be represented by a strain tensor ε and a stress tensor σ.

Fig. 5.18 shows the axes $0XYZ$ arbitrarily chosen within the sample and a generic direction, defined by the φ and ψ angles, along which the strain $\varepsilon_{\varphi\psi}$ is measured, with direction cosines $(\alpha_1, \alpha_2, \alpha_3)$ given by

$$\begin{aligned} \alpha_1 &= \sin\psi\cos\varphi \\ \alpha_2 &= \sin\psi\sin\varphi \\ \alpha_3 &= \cos\psi \end{aligned} \qquad (5.81)$$

The strain $\varepsilon_{\varphi\psi}$ is given by

$$\varepsilon_{\varphi\psi} = \alpha_1^2\varepsilon_{xx} + \alpha_2^2\varepsilon_{yy} + \alpha_3^2\varepsilon_{zz} + 2\alpha_1\alpha_2\varepsilon_{xy} + 2\alpha_2\alpha_3\varepsilon_{yz} + 2\alpha_3\alpha_1\varepsilon_{zx} \qquad (5.82)$$

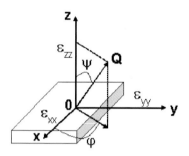

Fig. 5.18. Cartesian axes $OXYZ$ arbitrarily chosen within the sample and a generic direction, defined by the φ and ψ angles, along which the strain $\varepsilon_{\varphi\psi}$ is measured

This is a linear equation with six unknowns, ε_{xx}, ε_{yy}, ε_{zz}, ε_{xy}, ε_{yz}, ε_{zx}, which can be solved exactly, in principle, if the strain is measured along six independent directions. However, in practice, the accuracy can be improved if more directions are considered. The simple test directions are along, and at 45° to the axes OX, OY, OZ. After determination of the six unknowns. i.e. of the strain tensor, the principal strain axes $OX'Y'Z'$ and the principal strains ε along them may be found by diagonalization. The next section will show how the strain $\varepsilon_{\varphi\psi}$ can be measured in each chosen direction by using X-ray or neutron diffraction.

5.3.5 Principle of Strain Determination

The idea of strain measurement by neutron diffraction is shown in Fig. 5.19. In a typical neutron diffraction experiment [11, 115, 184] a collimated neutron beam, of wavelength λ, is diffracted by the polycrystalline sample (with a take-off angle of 2θ), passes through a second collimator and reaches the detector. The slits of the two collimators define the 'gauge' volume, the cross-section of which can be as small as 1 mm×1 mm and, in special cases, even smaller. The method of strain measurements by X-ray diffraction is similar. However the X-rays are less penetrating than neutrons; on the other hand the spatial resolution can be much higher when using Synchrotron Radiation namely of the order of a few tens of micrometres.

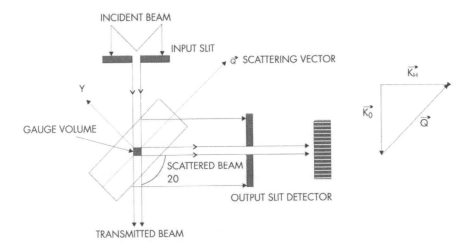

Fig. 5.19. Schematic representation of the strain measurements by neutron diffraction

The interplanar distance $d_{<hkl>}$ (where $< hkl >$ are the Miller indices of the investigated lattice planes) can be evaluated using the *Bragg law*

$$\lambda = 2d_{<hkl>} \sin \theta \tag{5.83}$$

(The 2θ take-off angle corresponds to the maximum of the Bragg-diffracted intensity peak.) The corresponding lattice strain is defined as

$$\varepsilon_{<hkl>} = \frac{d_{<hkl>} - d_0}{d_0} \tag{5.84}$$

where d_0 is the $< hkl >$ -interplanar distance in a stress-free material.

In practice the technique consists of precisely measuring the angular position of the Bragg peak for the stress-free and stressed material. From the observed shift one can obtain the value of the lattice strain and, by knowing the elastic constants of the medium, the value of the lattice stress, as explained above. Figure 5.20 reports the shift of the Bragg peak towards lower angular values in a region with tensile strains (as deduced from (5.83) where λ is constant) and towards higher angular values where the strains are of compressive nature.

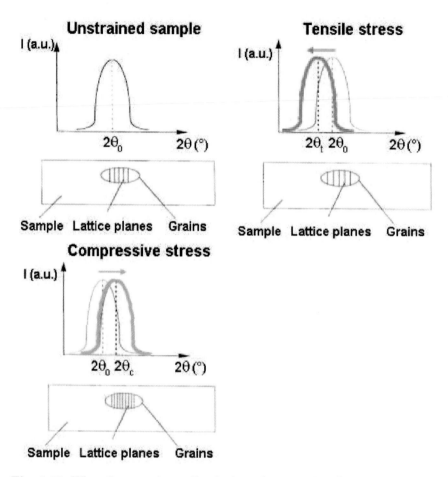

Fig. 5.20. When the stress is tensile, the interplanar spacing $d_{<hkl>}$ increases and the Bragg peak shifts towards smaller angles. The reverse occurs for compressive stresses

It must be emphasized that the direction of the measured strain is normal to the diffracting planes and is given by the $\mathbf{Q}$ vector in Fig. 5.19 ($\mathbf{Q} = \mathbf{k_H} - \mathbf{k_O}$) where $\mathbf{k_H}$ and $\mathbf{k_O}$ are the wavevectors of the diffracted and incident beams respectively). The same principle and equations are also valid for strain determination by X-ray diffraction. Relevant examples of the application of these techniques can be found in [152, 63, 237, 185].

5.3.6 Experimental Facilities

Neutron beams can be classified into two general kinds, which in general are linked to the neutron-generation method: continuous beams, which are generated by reactors, and pulsed beams, which are mainly found in spallation sources.

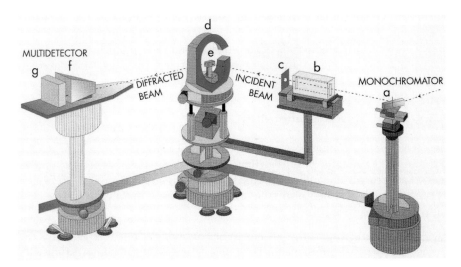

Fig. 5.21. The neutron diffractometer DIANE for residual stresses, installed at Laboratoire Léon-Brillouin, Saclay, France

Figure 5.21 shows a typical instrument dedicated to residual stress measurements: the DIANE diffractometer, using the continuous beam from the Orphée reactor installed at the Laboratoire Léon Brillouin (LLB) in Saclay (France), in the framework of a Franco–Italian (INFM) cooperation [33]. This instrument is equipped with a Euler cradle, allowing orientation of the sample automatically over large angular ranges. The working principle of X-ray diffractometers, including those installed at the Synchrotron Radiation Facilities are similar. The description of a sophisticated X-ray diffractometer for measuring residual stresses in the intermediate zone between surface and volume (depth between 10μm and 1 mm) can be found in [87].

5.3.7 The Stress-free Reference Sample

From (5.84) it appears that in order to determine the strain at a given point of a sample, the knowledge of the interplanar distance d_0 of the stress-free sample is necessary. Each error in d_0 implies systematic errors in the strain value. As a consequence it is necessary to determine d_0 with the highest possible accuracy. Several methods are commonly used to determine d_0. The most usual one is to measure the interplanar distance on a small sample of the same material submitted to a proper thermal treatment (annealing), in order to relieve any residual stress. A second method consists in measuring the interplanar spacing in a region of the investigated component, which is assumed to be stress-free because it is unaffected by the process inducing the residual stresses. A third method consists in measuring d_0, in an annealed powder of the investigated material, in which the assumption that the powder granulates are small enough that the stresses are relaxed, is reinforced by the annealing treatment. A fourth method makes use of the classical equilibrium equation, which states that the integrals of the stresses and their moments on any sample volume should be zero. In this case the d_0 value is obtained by an iterative procedure.

5.3.8 Metal Matrix Composites

As *Metal Matrix Composites* (MMC) are physically inhomogeneous, stresses are almost never uniformly distributed throughout their structure: applied loads tend to be unequally distributed between matrix and reinforcement because of the stiffness mismatch. Residual stresses in MMCs can be split into three categories [76]:

- A *Macrostress* (σ_{macro}), that has to be considered a *type I stress*. This category of stress is the same for both phases and is of prime interest for what concerns the material mechanical properties for engineering applications.
- A *Microstress*, i.e. a *type II stress*, which can occur between the two phases within a MM composite. In this case the microstress can be split into two contributions:

  1. An *elastic mismatch microstress* (σ_{mE}), due to the difference between the elastic constants of the two phases. It represents the load transfer to the stiffer phase (ceramic reinforcement). It is directly proportional to the macroscopic stress field.

  2. A *thermal mismatch microstress* (σ_{mT}), due to a difference between the thermal expansion coefficients of the two phases. Plastic deformation in the matrix or cooling from high temperatures (for instance during production) will cause an average tensile stress in one phase (e.g. in the Al-based matrix) and a compressive stress in the other (reinforcement in the previous example). This kind of stress is expected to be isotropic (hydrostatic).

Thus the following vector relation holds for both matrix and reinforcement phases (i):

$$\sigma^i_{tot} = \sigma_{macro} + \sigma^i_{mE} + \sigma^i_{mT} \tag{5.85}$$

The macrostresses can be calculated from the measured stresses in both phases as follows:

$$\sigma_{macro} = f\sigma^{reinf}_{tot} + (1 - f)\sigma^{matrix}_{tot} \tag{5.86}$$

where f is the volume fraction of the reinforcement.

The elastic mismatch microstress is related to the macrostress by means of the following equation:

$$\sigma^i_{mE} = \mathbf{B}^i \sigma_{macro} \tag{5.87}$$

where $\mathbf{B}^i$ is a fourth-rank tensor depending on the reinforcement particle shape and on the elastic constants of the two phases. It has been calculated and tabulated in literature [76] for different systems, on the basis of *Eshelby's* "equivalent homogeneous inclusion" *model* [68]. Finally, the thermal mismatch microstress in each phase can be calculated by means of (5.85).

For MMCs, the "wavelength" of the local stress fluctuation within the matrix is much smaller than the spatial resolution of the diffraction technique. Consequently, such stress variations cause a broadening of the diffraction peak. However, a non-zero average stress in each phase will give rise to a small, but measurable, shift in the position of the peak corresponding to that particular phase. An excessive load transfer should be responsible for the reinforcement-particles failure.

5.3.9 AA359 + 20 vol.% SiC$_p$ Brake Drum

In order to predict the in-service mechanical performances, residual stresses were investigated before and after bench tests on a AA359 + 20 vol.% SiC$_p$ brake drum prototype (Fig. 5.22) [25].

Fig. 5.22. AA359 + 20 vol.% SiC$_p$ brake drum

All the studied brake drums have been submitted to T6 treatment. This process could introduce some stresses (especially close to the surface) that are added to those introduced during the component processing and to those arising after mechanical treatments (fatigue-cycles).

Neutron diffraction was performed to evaluate the strain and stress state in the AA359+20%SiC$_p$ brake drum:

1. before bench test;
2. after submission to the following set of cycles without breaking:

   15000 N for 2065000 cycles;
   25000 N for 2600000 cycles;
   30000 N for 2500000 cycles;
   35000 N for 2500000 cycles.

3. broken after submission to 782000 cycles at 25000 N.

The experimental data were analysed through the previously explained method and the total stresses $< \sigma_{tot} >$ both in the Al-based matrix and in the SiC particles were obtained. It has to be recalled that such $< \sigma_{tot} >$ values are the average of the values obtained in the crystallites inside the gauge volume corresponding to the particular experimental set-up. The most interesting result was the macrostress monitoring in the wall region indicated in Fig. 5.23.

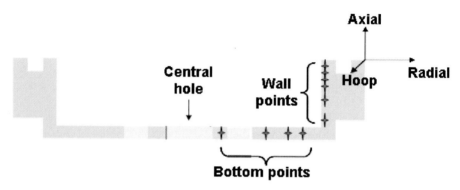

Fig. 5.23. Component geometry and measurement points

In fact the macrostress influences the mechanical performance of the in-service component. Macrostress obtained before bench tests are compared with those obtained after the two types of fatigue cycles (Fig. 5.24).

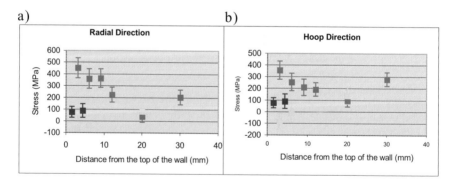

Fig. 5.24. Macrostress in the radial (a) and hoop (b) directions before bench tests (blue points), after bench test without breaking (pink points) and after bench test with breaking (yellow points)

Figure 5.24 shows that the residual macrostresses in the wall regions tend to increase after the set of fatigue cycles without breaking. In fact they sensibly increase after the fatigue cycles [25], releasing in the broken component and also going into compression in certain regions.

5.3.10 MMC's Drive Shaft for Helicopter

Another investigation of MMCs concerns the *residual stress* determination in a tube having a thickness of 19 mm, consisting of a matrix of the Al alloy AA2009 reinforced with 25% SiC particles and simulating a drive shaft for a helicopter [25].

Figure 5.25 reports the axial stresses determined by neutron diffraction, in the Al matrix and in the SiC reinforcement, respectively, after extrusion and cooling at room temperature. Compressive stresses are observed in the SiC reinforcement and tensile stress in the Al matrix. This is explained by the larger thermal expansion coefficient of the Al alloy, which shrinks, during cooling, more than the SiC particles, which become compressed. According to Newton's third law, the matrix finishes under tension.

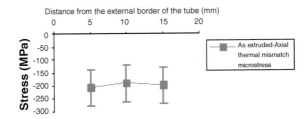

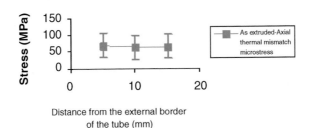

Fig. 5.25. Thermal mismatch microstresses in the axial direction in the SiC reinforcement (a) and Al matrix (b)

5.3.11 Welds

During the welding process, very strong temperature gradients occur, which induce very high residual stresses, sometimes of the order of the magnitude of the yield stress. It is therefore of great interest to determine experimentally these stresses. The neutron diffraction technique and the synchrotron radiation were used to investigate the strain profiles close to the welds. A representative example is reported here.

Al alloys for aerospace technology

Residual stresses induced by *Variable Polarity Plasma Arc* (VPPA) welding in an Al alloy 2219-T851 plate (6.5 mm thickness, 62 mm length in the direction parallel to the weld axis, 48 mm in the direction perpendicular to it) were investigated by neutron diffraction. This alloy is used in the spacecraft industry, as a major structural material for the NASA Space Shuttle and fuel tanks and for the ESA rocket Ariane 5 and the Alpha Space Station [9].

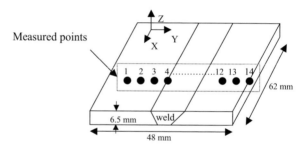

Fig. 5.26. Geometry of the Al 2219-T851 welded plate for spacecraft industry

Results in Fig. 5.27 show that the parent material is almost stress free. In the weld pool, tensile bulk (around 60 MPa) stresses were found, whereas the opposite occurs in the *heat-affected zones* (HAZ).

Curve shapes are consistent with theoretical and experimental behaviours reported in literature, and suggest the presence of "structure change stresses", according to Macherauch's definition. This means that a superposition occurs of the shrinkage stress due to different cooling rates inside and at the borders of the weld, and structural change stresses due to the presence of different phases. Processing history principally determines the residual stress state. Mechanical stirring changes the microstructure considerably but has only a secondary effect on the stress field.

5.3.12 Cold Expanded Hole

The simplest way to *arrest crack propagation* in metal alloys is the use of the cold expansion method at the crack tip. A decrease in the notch sharp-

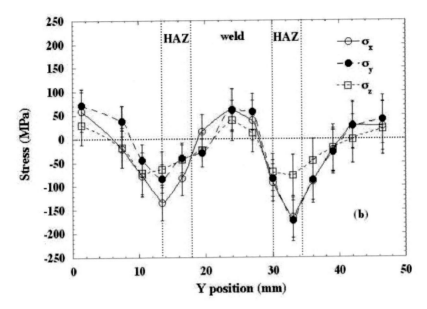

Fig. 5.27. Residual stress profiles across the Y direction of the weld, as obtained by neutron diffraction

ness is generated by drilling at the crack tip, and the residual compressive stresses are generated by a cold expansion. However, the effect is limited by the hole size, influencing the structure strength. By practising cold expansion, the lifetime is strongly improved and the minimum crack propagation rate is lower. More effective results can be obtained by expanding a hole at the crack tip so as to produce residual compressive stresses that can reduce the effective stresses around the crack tip. The knowledge of the residual stress field in the neighbourhood of the expanded hole is fundamental to verify of the effectiveness of the method, and also for the lifetime prediction of the components submitted to it. Neutron diffraction was applied to determine residual stresses in a specimen of A42 ferritic steel, and the results are compared with finite element (FEM) calculations [31]. The specimen, obtained by lamination, was pre-cracked with a 27.5 mm long crack. A hole (diameter 5.9 mm) was drilled at the crack tip, and subsequently cold expanded using a 6 mm diameter steel ball. The specimen was then submitted to fatigue cycling (frequency 30 Hz, stress amplitude 3.75 kN, maximum cyclic stress 44 MPa). Experimental results are shown in Fig. 5.28a, compared to FEM calculations (Fig. 5.28b). These results are in satisfactory agreement. In particular, the measured compressive minimum (about–300 MPa) of the z component near the hole is confirmed by FEM calculations, and so is the general behaviour in the x direction, at least in the neighbourhood of the hole. Higher compressive stresses are obtained from experiments with respect to FEM calculations in

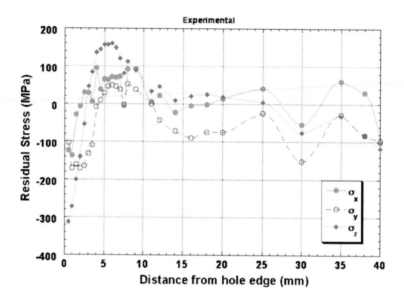

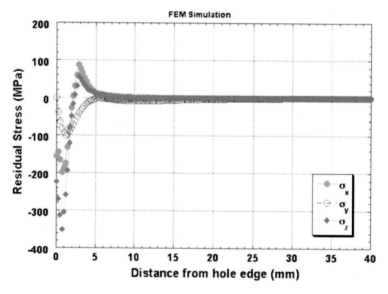

Fig. 5.28. Residual stress distribution in the neighbourhood of the expanded hole in pre-cracked A42 ferritic steel, as determined by neutron diffraction experiments (a) and FEM calculations (b); x = normal to sample plane, y = parallel to crack direction, z = transversal to crack direction [31]

the y direction. This can be reasonably ascribed to already existing stress states, probably due to the lamination process, which are not taken into account in the FEM calculations.

5.3.13 Shot Peening

Shot peening is an effective method of improving the fatigue life of engineering components by introducing compressive residual surface stresses which can approach the yield stress of the material. Careful selection of the peening parameters is necessary to produce the optimal residual stress distribution for maximum fatigue life improvement.

Verification of the residual stresses introduced by shot peening is necessary. Carried out by X-ray diffraction it supposes removal of material. But using neutron diffraction or hard X-ray diffraction (synchrotron radiation) it becomes nondestructive and allows investigation of the bulk.

5.3.14 Residual Stress Study of Laser Shock Peening for Ti-6Al-4V Alloy for Aerospace Application

Laser Shock Peening (LSP) has been used on the root sections of Ti-6Al-4V (cf. Sect. 1.2) aerospace components to improve their fatigue performance (Fig. 5.29). Laser shock peening is capable of producing compressive resid-

Fig. 5.29. Aeroplane engine turbine

ual stresses to a much greater depth than the conventionally used technique of shot peening. Figure 5.30 reports the *residual stress* profiles in the three orthogonal directions; the Z direction perpendicular to the surface and the X and Y directions in the plane. These profiles were obtained by using high-energy synchrotron X-ray diffraction [127]. It appears that the residual stress perpendicular to the peened surface (Z) is slowly changing and relatively small.

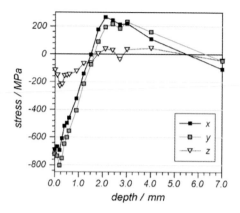

Fig. 5.30. Residual stresses profiles in a laser shot peened component of Ti-6Al-4V, as obtained by high-energy synchrotron X-ray diffraction

Laser shock peening is an effective technique to generate favourable compressive residual stresses in the near surface region of Ti-6Al-4V alloys. These stresses are expected to improve the material's resistance to fatigue cracking, allowing longer operating lifetimes for laser shock peened components.

5.3.15 Reed Pipe Brass Tongues of Historic Organs

True Baroque organ music can only come back to life in the 21st century by developing suitable Cu-based alloys and implementing them in the organ reed pipes. Reed pipes contain a vibrating part, the brass tongue that crucially influences its sound. In the frame of the European Truesound Project energy dispersive synchrotron X-ray diffraction has been performed in order to investigate *residual stresses* in historic tongues belonging to several European organs built during the 18th century (Fig. 5.31) [155]. The in-depth analysis gives us an important indication of the processes the tongues were submitted to during their manufacturing: hammering, annealing, filing to the neat thickness, curving of the tongues. A biaxial stress state in the organ tongues was considered. The residual stress values and behaviour were correlated to the manufacturing processes.

Five samples were analysed at the synchrotron BESSY II (Fig. 5.32). In general, compressive orthogonal stress (Fig. 5.31) were observed in all the samples, close to the surface of the tongues, due to the hammering process. Close to the sample surface, after filing treatments, all the samples exhibit, in the direction parallel to the process, a relaxation of the compressive stress induced by hammering. In four of the five samples, less compressive stresses were observed in the parallel direction with respect to the orthogonal one for both sides of the tongues, which means that the filing was performed both on the front and on the back of the tongues. In one case (Fig. 5.33), the

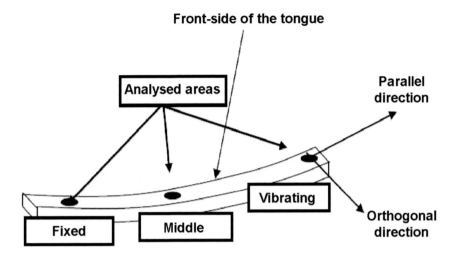

Fig. 5.31. Measurement points and directions in the investigation of organ tongues by X-ray diffraction

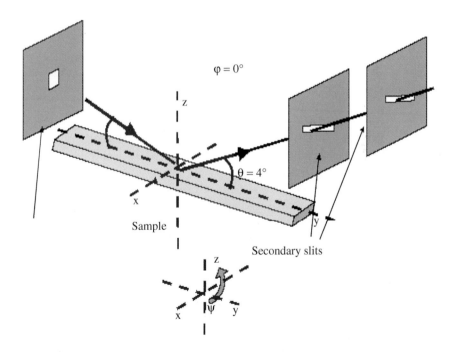

Fig. 5.32. Scheme of the multipurpose diffractometer for residual stress and texture analysis at BESSY II

stresses in the parallel direction on the back are similar to those obtained in the orthogonal direction, while those on the front are less compressive. This implies that for this sample, the filing was performed only on the front of the tongue.

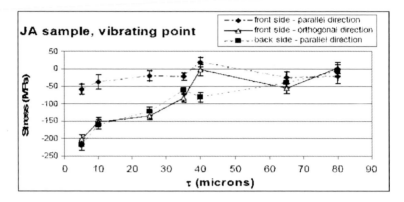

Fig. 5.33. Residual stresses in the vibrating area front and back in the parallel direction and in the orthogonal direction

As a consequence, the residual stress analysis performed on historic brass tongues allowed the authors to identify the main manufacturing processes the tongues were submitted to, giving precious information to the organ builders in their work of reproducing the sound of Baroque organs.

5.4 Three-dimensional Imaging by Microtomography of X-ray Synchrotron Radiation and Neutrons

A revolutionary discovery in the field of imaging occurred at the beginning of the 1970s when the first equipment for *Computed Tomography* (CT) was developed. This method of imaging avoids several important limitations of conventional X-ray or neutron radiography. CT avoids the superimposition inherent to radiographic imaging by producing slices in the third dimension in a nondestructive way with contrast discrimination up to one thousand times better than that of a conventional radiograph [48].

X-ray *microtomography* (microCT) is similar to conventional CT systems usually employed in medical diagnoses and industrial applied research. Unlike these systems, which typically have a maximum spatial resolution of about 0.5 mm, advanced microCT is capable of achieving a spatial resolution up to 0.3 μm [109], i.e. about three orders of magnitude better. Such a high spatial resolution can be obtained only for samples of reduced size, i.e. for dimensions in the range of a few cubic millimetres. *Synchrotron Radiation* (SR) allows quantitative high spatial resolution images to be generated with high signal-to-noise ratio [141, 170, 195]. Use of synchrotron X-rays has several advantages compared to laboratory or industrial X-ray sources. These include:

(i) a high photon flux which permits measurements at high spatial resolution;
(ii) the X-ray source is tunable, thus allowing measurements at different energies;
(iii) the X-ray radiation is monochromatic, which eliminates beam hardening effects; and
(iv) parallel beam acquisition allows the use of exact tomographic reconstruction algorithms.

In addition, SR allows acquisition of volumes at different energies and volume subtraction to enhance contrast. This variability in materials and acquisition methods leads to a wide range of greyscale values (corresponding to different X-ray absorption coefficients) among data sets [12].

Neutron Tomography (NT) is a technique used to derive plane or volumetric maps of the neutron attenuation coefficient values of an object. Thereby the morphology or structure of an object can be visualized nondestructively. The main difference between tomography with X-rays and neutrons is the following: X-rays interact with the atomic shell, i.e. they are scattered or absorbed by electrons. The more electrons an element has, the more it attenuates X-rays. Neutrons, on the other hand, interact with the atomic nuclei, but the absorption or scattering cross-sections do not show any obvious regularity across the periodic table of elements. Interaction strongly depends on the inner structure of the atomic nuclei, meaning that even isotopes of the same element may often provide very different levels of contrast in the projection. The high degree of neutron scattering caused by hydrogen and the penetration capacity of neutrons in most metals are of particular industrial significance [122].

The spatial resolution of the X-ray CT image is dependent on the number of parallel beam projections and the number of data points in each projection. A larger data set means a more detailed description of the depicted object and hence more pixels and of smaller dimensions, i.e. better spatial resolution. An important issue is the choice of spatial resolution versus overall sample size. Ideally, the specimen should absorb about 90% of the incident radiation along the most radio-opaque path to obtain the best signal to noise ratio in the reconstructed image. In a homogeneous sample, absorbing 90% of the incident radiation, the quantity $\mu(\lambda) x$ (where λ is the X-ray wavelength, $\mu(\lambda)$ is the linear attenuation coefficient of the sample for this wavelength, and x is the sample thickness), should be approximately 2. To satisfy this condition the sample thickness and/or the X-ray energy should be optimized. Neutron tomography enables investigations of the macroscopic inner structure of large samples (up to hundreds of cubic centimetres) with a spatial resolution of up to 100 μm. However, recent neutron tests show a spatial resolution as good as 50 μm [128]. In both three-dimensional (3D) microCT and NT, hundreds of two-dimensional (2D) projection radiographs are taken of the specimen at different rotation angles. A schematic of the set-up of a microCT system is illustrated in Fig. 5.34. If the sample is then imaged several times at different

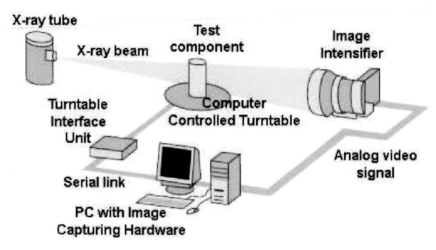

Fig. 5.34. Schematic set up of microCT system

orientations, 3D (volume) information on the sample structure can be obtained by using computer algorithms. This process, referred to as *tomographic image reconstruction*, consists in solving an inverse problem to estimate an image from its line integrals in different directions, in 2D, and the problem is theoretically equivalent to the inversion of the *Radon Transform* of the image. In practice, there are two major classes of reconstruction algorithms that use fundamentally different approaches to accomplish this conversion [178]:

(i) transform-based methods using analytic inversion formulae, and
(ii) series expansion methods based on linear algebra.

3D renderings of data after reconstruction may be made by electronically stacking up the slices. These 3D renderings may also be sectioned in arbitrary ways and could be planed, zoomed and rotated to better locate individual details. While the slice image and 3D renderings are very useful for making qualitative observations of an internal concrete structure, the real benefit is the quantitative information that can be extracted from the 3D data sets. Different methods may be applied to extract quantitative architectural parameters from the tomographic images, based on the structural indices usually measured for bone samples.

MicroCT and NT allow accurate non-destructive 3D examination of samples. The development of microCT was first driven by the need for having a highly precise means of reconstructing the complex architecture of bone tissue at a high resolution, and is becoming a critical parameter in the evaluation of disease pathogenesis and efficacy of interventions. Nowadays, there are numerous medical and nonmedical imaging applications of microCT.

5.4.1 Tissue Engineered Bones

One interesting application of X-ray microCT concerns the investigation of the structure of biomaterials and bone produced by stem cells (tissue engineering). Highly porous hydroxyapatite scaffolds were investigated before implantation and after they were loaded with *in vitro* expanded *Bone Marrow Stromal Cells* (BMSC) and implanted for 8, 16 and 24 weeks in immunodeficient mice [130]. Synchrotron X-ray computed microtomography (microCT) was used for three-dimensional (3D) characterization of the scaffold material and 3D evaluation of tissue engineered bone growth after *in vivo* implantation. Experiments were performed taking advantage of a dedicated set-up at the *European Synchrotron Radiation Facility* (ESRF, Grenoble), which allowed a spatial resolution of about 4.91 μm.

The growth kinetics was investigated by determining the volume fraction, the average thickness and distribution of the newly formed bone as a function of the implantation time. New bone thickness increased from the 8th to the 16th week, but deposition of new bone was arrested from the 16th to the 24th week. Instead mineralization of the newly deposited bone matrix continued up to the 24th week. Figure 5.35 shows in the panels a–c the 3D structure of scaffolds after implant at 8 (a), 16 (b) and 24 (c) weeks. The images show the new bone (green) on the inner surface of the scaffold (yellow). The other phases (e.g. organic phase) are blue. Panels d–e show a 3D display of the new bone (d), and a zoom of the area of interest (e). Figure 5.36 reports the evolution of the new engineered bone thickness as a function of implantation time.

5.4.2 Al+4% ZS Composites

High resolution synchrotron X-ray microtomography was used to image damage initiation and development during mechanical loading of structural metallic materials [27]. First, the initiation, growth and coalescence of porosities in the bulk of two metal matrix composites have been imaged at different stages of a tensile test. Quantitative data on *damage development* has been obtained and related to the nature of the composite matrix. Moreover three-dimensional images of *fatigue cracks* have been obtained *in situ* for two different Al alloys submitted to uniaxial external load. The analysis of these images shows the strong interaction of the cracks with the local microstructure and provides unique experimental data for modelling the behaviour of such short cracks.

The investigated sample were two different *Al-based metal matrix composites* reinforced by spherical *zirconia based particles* with a size ranging between 40 and 60 μm (4% volume fraction). Figure 5.37 reports the scheme of the device used to perform X-ray microtomography experiments *in situ* on the samples

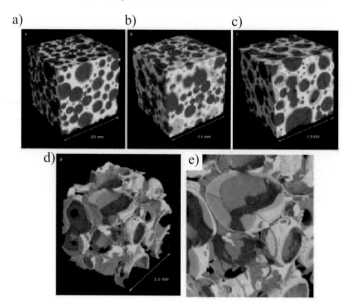

Fig. 5.35. Panels a–c: 3D display of subvolume of scaffolds after implant for 8 (a), 16 (b) and 24 (c) weeks. The images show the new bone (green) on the inner surface of the scaffold (yellow). The other phases (e.g. soft tissues) are blue. Panels d–e: 3D display of the new bone (d), zoom of the area of interest (e)

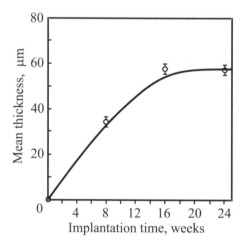

Fig. 5.36. Evolution of the new bone thickness as a function of the implantation time (growth kinetics)

subjected to monotonic load. Figure 5.38 reports the reconstructed internal sections, obtained by 3D images, of the Al+4% ZS composite damage (a) and of the Al 2124 (T6) +4% ZS. It appears that under the external load in the first sample the damage occurs mainly by *decohesion* at the *particle/matrix* interface, whereas in the second sample the damage occurs by zirconia *particle cracking*. The observed differences, in terms of damage events, seem to be closely linked with the different yield stresses of both matrix. Other experi-

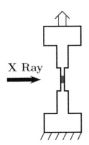

Fig. 5.37. Scheme of the *in situ* 3D imaging of composite samples under external load

mental data can be found in [27] including imaging of samples submitted to fretting and to uniaxial *in situ* fatigue.

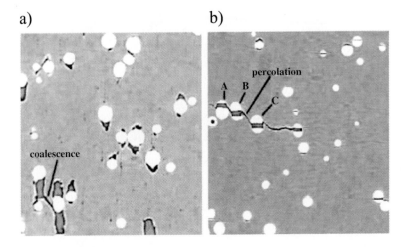

Fig. 5.38. Reconstructed images of the interior of the Al/ZrO$_2$ composites: (a) Al + 4% ZS, main damage mechanism: particle/matrix decohesion and (b) Al2124(T6) + 4% ZS, main damage mechanism: particle cracking

5.4.3 Sintering of Cu Particles

Another example concerning material science is given by the investigation by X-ray microCT of the microstructural evolution at the 3D level of copper spherical particles submitted to a sintering process [139]. Figures 5.39 and 5.40 reports the microstructure evolution of the sample during the sintering process. Bidimensional cross-sections perpendicular to the axis of the cylinder were obtained from the reconstructed 3D images. The results of this nondestructive kinetic investigation will be used in future models of the sintering process. Bidimensional cross-sections perpendicular to the axis of the cylinder

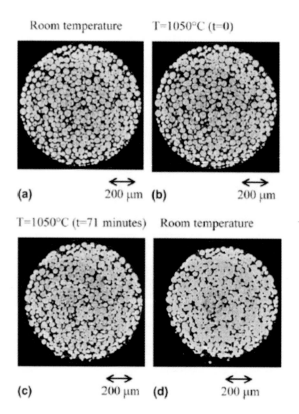

Fig. 5.39. 2D reconstructions (virtual slices) perpendicular to the cylindrical axis showing Cu particles at different stages of the sintering process

were obtained from the reconstructed 3D images. The results of this nondestructive kinetic investigation will be used in future models of the sintering process. Concerning neutron tomography the neutron beam can pass through centimetres of metal but it is easily attenuated by small amounts of light elements like hydrogen, boron and lithium. This makes neutron tomography unique tool for *nondestructive testing* with applications in industry, material science and various other fields [57] and [197].

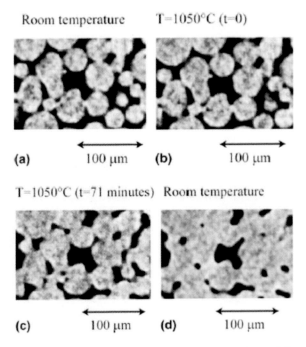

Fig. 5.40. Magnified parts of the reconstructed slices of Fig. 5.39

5.4.4 Al Foams

In the framework of an international collaboration, open and closed cell *Al foams* with increased vibration damping performances are being designed. The followed route includes modelling by theoretical and computational methods, and subsequent processing, performed by the Fraunhofer Institute for Manufacturing Technology and Applied Materials Research (IFAM) - Bremen, by means of its own property technique called *Advanced Pore Morphology* technology (APM).

The general concept of APM is the separation of two main process steps:

(i) foam expansion;
(ii) foam part shaping.

APM foam parts consist of small volume metallic foam elements, which are expanded in bulk/mass production. Joined to each other in a separate process step, the foam elements finally form the APM foam part. The density obtained is in the range 0.3 – 1 g/cm^3.

The efficient use of foams requires a detailed description of their structure. In fact, the foam performances are known to be dependent on density and cell geometry, as well as on the properties of the solid material from which the cell face is made. The geometrical features include cell shape and size, distribution in cell size, defects and flaws in the cell structure as well as microstructural parameter like cell wall and edge thickness and material distribution between cell face and cell edge. Therefore a complete characterization of the 3D structure of the material is required.

Figure 5.41 reports the images obtained on an Al foam sample using phase contrast neutron tomography [29]. The 2D sections were deduced from the 3D microstructure. The results obtained will be used in the modelling and in the

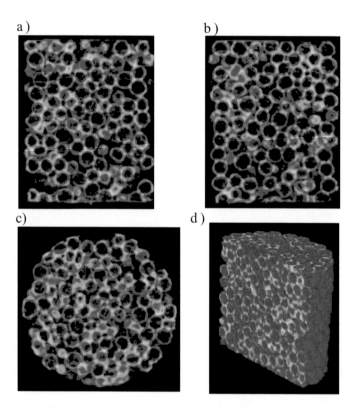

Fig. 5.41. The 2D sections obtained by phase contrast neutron tomography of an Al foam sample [29]: (a) xz plane frontal section, (b) yz plane sagittal section, (c) xy plane – axial section 70 slices more inside the sample, (d) right section of the sample

optimization of the parameters determining the foam properties, such as the element density, element geometry and volume, type of adhesive, thickness of adhesive coating, etc.

The microCT techniques have found several other applications: in cement and concrete research they have produced fruitful results in the study of alkali-silica reaction [21], and encapsulation of hazardous waste materials [28]. Other examples are investigations concerning quality tests of soot filters, adhesive joints, lubrication films and *in situ* visualization of water management in fuel cells. Interesting applications of neutron tomography in archaeology [149] and biology [236] were recently reported. Ref. [124] reports a review of industrial applications of neutron tomography.

5.5 Specific Investigations of Functionally Graded Materials (FGM) Systems for High Temperature Applications

Materials for advanced high temperature structural applications, including spacecraft heat shields, coatings on gas turbine blades, plasma facings for fusion reactors, are required to have strength at high temperatures, creep resistance, adequate toughness and thermal shock resistance. While ceramics have low density, creep resistance and good high temperature strength, their fracture toughness and thermal shock resistance are poor, limiting their use in demanding applications. On the other hand, metallic materials have high fracture toughness and excellent thermal shock resistance while their weak point is the poor strength-related properties at high temperature (cf. Sect. 1.2). Joining ceramics to metals to combine the advantages of these two kinds of materials has proved to meet the material requirements in many applications. However, thermal residual stresses arising from cooling after high temperature processing and problems due to poor chemical compatibility limit the wide use of ceramic–metal joints. One of the main problems has been the large difference in thermal expansion coefficients of ceramics and metals that are of interest to structural applications (cf. Sect. 1.2). The stresses that arise after joining many ceramic–metal combinations exceed the fracture strength of ceramic, especially in regions close to the free surface near the interface. The result is either cracking of the ceramic or poor strength of the joints.

Functionally Graded Materials (FGM) were introduced to reduce the stresses due to the mismatch in the thermal expansion coefficients between the ceramic and metal parts. FGMs are usually a combination of two or more material phases that have a gradual transition from one material at one surface to another material at the opposite surface. In a metal–ceramic FGM, the metal-rich side is situated in the region where mechanical performance is needed, while the ceramic-rich side is exposed to high temperatures or placed in regions where extreme temperature variations occur. The intermediate regions consist of a mixture of the constituents, varying in volume fraction with distance. This means a gradual change in thermal expansion coefficients, minimizing in this way the thermal stresses arising from cooling or heating. Moreover, metallic phases embedded in ceramic could increase the thermal conductivity, reducing the temperature gradient across the thickness and hence minimizing the susceptibility to thermal shock-induced fracture.

The presence of local heterogeneous microstructure may cause locally concentrated residual stresses during the thermal or mechanical loading. These locally concentrated stresses, especially those high in tension, can act to initiate small cracks and voids. The development of these small-scale failures may lead to large-scale failures and result in fracture of the whole structure.

For these reasons it is extremely important to analyse both the microstructure and the residual stresses present in the FGM systems for high temperature applications.

The *residual stress* state in a FGM coating can be analysed using both neutron and synchrotron radiation diffraction. Neutron diffraction is suitable in the case of thicker layers, while synchrotron radiation is used for thinner coatings because a much higher spatial resolution can be obtained due to the possibility to limit the beam width by using slits and still have sufficient beam intensity.

5.5.1 Residual Stresses in W/Cu FMG

Residual stresses were analysed in W/Cu FGM for application as part of a fusion reactor wall or as a high current switch (cf. Table 1.7). The composites produced were free of porosity and had an interpenetrating network microstructure. The residual stresses were evaluated by neutron diffraction in steps along the concentration gradient, averaging over the cross-section with a beam width of 1 mm^2 (Fig. 5.42). The observed stresses often exhibit much

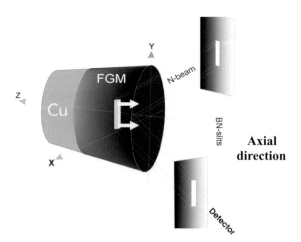

Fig. 5.42. Scheme for neutron stress analyses with FGM. The white pointer represent the scattering vector and the strain direction of that measurement

higher values of opposite sign than expected taking into account only thermally induced stresses. These results demonstrate that nonequilibrium effects like inhomogeneous and fast cooling, plastic Cu deformation, and microcracks in the W-matrix determine micro- as well as macrostresses. After an appropriate thermal treatment the detected distribution of macrostresses is in better agreement with FEM calculations [199].

5.5.2 Neutron Strain Mapping within Ni/Al$_2$O$_3$

Neutron strain mapping within a Ni/Al$_2$O$_3$ functionally graded specimen was performed on the alumina end and into the alumina–nickel composite interlayer (Fig. 5.43) (cf. also Table 1.5). Overlapping sampling volume elements

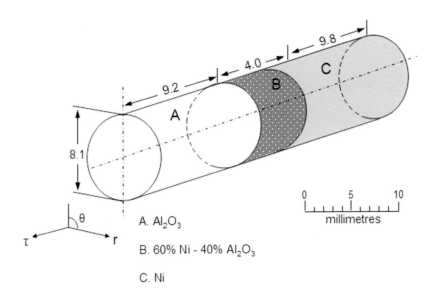

A. Al$_2$O$_3$

B. 60% Ni - 40% Al$_2$O$_3$

C. Ni

Fig. 5.43. Scheme of the analysed Ni/Al$_2$O$_3$ functionally gradient specimen

aided in characterizing the steep strain gradients. Clear evidence of steep residual stress gradient suggested by the strain gradient near the interface between the Al$_2$O$_3$ layer and the alumina–nickel composite interlayer was obtained (Fig. 5.44) [186]. The continuous line was obtained by finite element calculations and is in excellent agreement with the experimental data. At the interface an opposite sign behaviour is evident, as usually observed at the metal–ceramic interface: the stresses are compressive in the ceramic and, for Newton's third law, tensile in the metal.

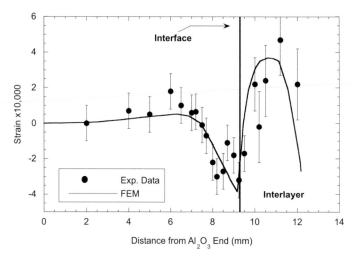

Fig. 5.44. Strain values in the Al₂O₃ layer and the alumina–nickel composite interlayer, in the direction of the cylinder axis

5.5.3 Residual Stresses in Ni/8Y–ZrO₂ FGM

In comparison to neutron diffraction, high energy synchrotron radiation offers the possibility to analyse the stress state within a considerably smaller gauge volume, thus increasing the spatial resolution. Therefore, it is suited to the evaluation of the residual stress distributions in FGMs, even if they contain steep compositional gradients [53]. A microwave sintered Ni/8Y–ZrO₂ FGM with a thickness of 6 mm consisting of 11 individual layers with 0, 10, 20,...,100 vol.% Ni was analysed with regard to the phase-specific residual stresses within both the metal and the ceramic phase. Due to the lower thermal expansion coefficients of the ceramic phase, compressive residual stresses were determined in the zirconia. They are partially compensated by tensile residual stresses in the ductile metal component. On the macroscopic scale two local maxima of compressive residual stresses, one in the ceramic-rich area and one in the metal-rich area are observed (Fig. 5.45). Compared to neutron diffraction methods, considerably shorter exposure times as well as an improved spatial resolution are achieved in this experiment performed using synchrotron radiation.

5.5.4 Residual Stresses in WC/Co FGM

X-ray diffraction was used to determine the *thermal residual stresses* that develop in a functionally graded WC/Co composite. WC/Co composites are commonly used as tool bits, where the presence of Co improves the toughness of the wear resistant WC [143]. In order to obtain better overall functionality in these composites, WC and Co phase volume fraction gradients are

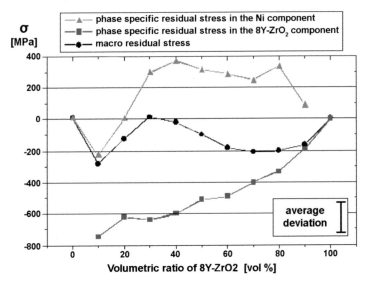

Fig. 5.45. Phase specific and macro residual stresses in a Ni/8Y–ZrO$_2$ FGM as a function of the volume fraction

introduced during the process. Thermal residual stresses develop due to the thermal mismatch between the WC and the Co phase during cooling from the sintering temperature and can hence be expected to influence their mechanical properties.

The gradient in the specimen was continuous within a distance of about 40 μm below the surface and consisted of varying WC and Co phase volume fractions. The stresses were determined at various depths in the functionally graded material. The compressive residual microstresses in WC increase almost linearly, from approximately –80 to –170 MPa, with increasing Co content. For the low volume fraction Co phase, the tensile residual stresses were approximately 600 MPa. The changes in the magnitude of the residual stresses were related to the changes in the Co phase content through the graded zone. Microstresses in the graded zone are attributed to the thermal mismatch between WC and Co phase. The compressive macrostresses are a result of the compositional gradient. This conclusion is supported by the fact that almost no macrostresses were measured in a similar homogenous sample, i.e. without the corresponding compositional gradient.

5.5.5 Small Angle Scattering Experiments

The microstructure of the FGMs can be nondestructively analysed using the small angle scattering technique or by performing a microtomography experiment.

As mentioned in Sect. 5.3 an important technique is *Small Angle X-ray Scattering* (SAXS) using high-energy X-rays. SAXS is widely used in materials science to analyse inhomogeneities ranging in size from approximately 0.001–1 μm, and thus offering complementary information to the usually destructive techniques for morphological characterization, such as TEM. High-energy X-rays provide deep penetration in a wide range of materials, being particularly well suited for studies requiring high spatial and/or temporal resolution.

SAXS can be used in order to analyse functionally graded materials for *thermal barrier coatings* applications. In these materials, spatial variations occur on the scale of tens of microns, and gradients in microstructure must be correlated with those in phase at identical locations. Thus, identical, undisturbed scattering volumes are needed for morphology characterization by SAXS and phase characterization by wide angle X-ray scattering (WAXS). In addition, the microstructures in these systems are typically directional, so 2-D data is needed to determine this orientation dependence. This combination of needs, previously unavailable by existing techniques, is well addressed by the HE-SAXS and HE-WAXS instruments of the Advanced Photon Source at the Argonne National Laboratory (USA), which can be employed simultaneously.

5.5.6 SAXS on a Thermal Barrier Coating

A thermal barrier coating (TBC) system was analysed, consisting of an 800 μm thick 78 wt.% Y_2O_3-stabilized ZrO_2 coating (topcoat) on a 50 μm thick Ni–Co–Cr–Al–Y coating (bond-coat) on a stainless steel substrate [92]. The coatings were deposited using electron-beam deposition. The SAXS profiles show strong directionality and vary strongly through the topcoat and into the bondcoat. Detailed 2-D analysis indicates the presence of a hierarchy of pores, ranging from globular pores to featherlike cracks, with the overall amount of nanoscale porosity decreasing with increasing coating thickness. The WAXS measurements over the same regions clarified the presence of gradients in crystallographic texture and phase composition. The latter result was used to quantify complementary radiography measurements and provide a second measure of depth-resolved porosity.

5.5.7 Residual Stresses and SANS on Al_2O_3/Y–ZrO_2 FGM

A complementary residual stress and small angle scattering analysis (SANS) was performed on new functionally graded Al_2O_3/Y–ZrO_2 ceramics [154] (cf. also Table 1.1). A neutron diffraction mapping of residual stresses was done in order to correlate the mechanical properties to the distribution of residual stresses resulting mainly from phase specific stresses after cooling from the sintering temperature. In addition, the SANS technique was applied to study the porosity in the ceramics as a function of the production parameters. The influence of *Hot Isostatic Pressing* (HIP) on the porosity of the final product was examined.

It was found that in the case of the sintered and HIPed FGM specimen, the stress in the alumina phase is of a compressive nature and about 130 MPa

in the Al_2O_3/ZrO_2 (80/20) core. According to the equilibrium condition the
calculated stress in the zirconia phase is a tensile stress of +520 MPa. A significantly lower level of interphase stresses was detected in the core of the
sintered FGM disc prior to HIPing. In this case, the stress in the alumina
phase is determined to be −90 MPa, whereas the stress in the zirconia phase
is +360 MPa.

As regarding the SANS results, a very significant effect on porosity reduction
was observed when applying a post-HIP treatment. This process reduces the
original 20.3% porosity after HIP. Moreover, a very positive role in this process could be ascribed to the presence of $Y-ZrO_2$ grains in the phase mixture
(Fig. 5.46). The most important conclusions could be made when combining results of both neutron scattering techniques. A significantly higher level
of phase specific stresses have been found in specimens with a very low volume fraction of pores. This indicates a relatively strong relaxation of internal
stresses in dual phase ceramics due to porosity.

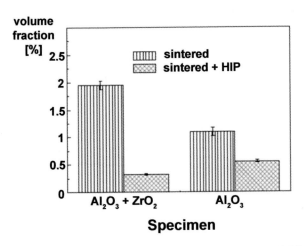

Fig. 5.46. Comparison of pore volume fractions in sintered and HIPed ceramics

5.5.8 MicroCT on Al/SiC_p FGM

In the work [224] microCT was used for the elucidation of some aspects regarding particle distribution in SiC-particle-reinforced functionally graded aluminium (FGM) composites. In particular, the data allows one to obtain 3D
qualitative information about pore and particle distributions (Fig. 5.47). It
can be observed that as a consequence of centrifugal casting, SiC particles
are partially clustered, with some pores gathered with them, this being associated with imperfect wetting of ceramic particles by the molten aluminium
alloy. By comparing the shape of the particles in the reconstructed images to

that obtained by SEM, it should be remarked that the reconstructed particles exhibit much smoother edges.

Also, small spherical pores, associated to trapped gas bubbles, are observed dispersed in the matrix and it was possible to quantify the particle volume fraction profile along the centrifugally cast FGM composite. In addition, a relatively smooth gradient of ceramic particles was observed along the sample, which can be beneficial in terms of mechanical performance of the material.

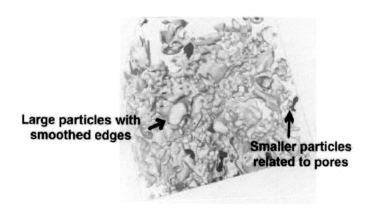

Fig. 5.47. Simultaneous 3-D rendering of pores and particles in Al/SiC$_p$ FGM

Appendix: European Sources of Neutrons and X-ray Synchrotron Radiation and their Main Instrumentation

A.1 Neutron Sources and Laboratories

A.1.1 ILL Grenoble (France)

The Institut Max von Laue-Paul Langevin (ILL) was founded in January 1967 on the initiative of France and Germany, with the United Kingdom taking an equal share in 1973. The aim of this enterprise was to create an intense source of neutrons entirely dedicated to civil fundamental research. Later on ten more countries were associated to ILL. Neutrons at the ILL are produced by its High Flux Reactor. Refurbished between 1993 and 1995, it produces the most intense neutron flux in the world: 1.5×10^{15} neutrons/(second·cm^2), with a thermal power of 58.3 MW. At the beginning of the third millennium, an instrument renewal action was undertaken (the so-called "Millenium Programme") to ensure that ILL remains at the forefront of modern science. Details on the available instrumentation at ILL can be found in [104].

A.1.2 ISIS (UK)

ISIS, at Rutherford Appleton Laboratory, is a pulsed source: a heavy metal target is bombarded with pulses of highly energetic protons from a powerful accelerator, driving neutrons out from the nuclei of the target atoms (the process is called "spallation"), thus realizing a pulsed neutron source. In contrast with reactor sources of neutrons, suffering from intense heat production in the core, the low duty cycle of the ISIS accelerator corresponds to a time averaged heat production at the target of only 160 kW. In the pulse, however, the neutron brightness exceeds that of the most advanced steady state sources. Thus an extremely intense neutron pulse is delivered with only modest heat production in the neutron target. ISIS supports an international community of around 1600 scientists who use neutrons and muons for research in physics, chemistry, materials science, geology, engineering and biology. Details on the instrumentation available at ISIS can be found in [105].

A.1.3 Laboratoire Léon Brillouin (LLB) Saclay (France)

The scientific activities of the Laboratoire Léon Brillouin (CEA-CNRS) in Saclay (F) can be classified into three fields of equal importance: physical-chemistry, structural and phase transition studies, magnetism and supercon-ductivity. The LLB is a service-providing laboratory that also develops its own scientific research. One of its roles is the design, construction, and oper-ation of neutron spectrometers installed around the Orphée reactor in Saclay. These are mainly used for the study of condensed matter. Details on the instrumentation available at LLB can be found in [106].

A.1.4 BENSC, Berlin (Germany)

The Berlin Neutron Scattering Centre (BENSC) is hosted by the Hahn-Meitner Institute (HMI) of Berlin (D). The latter is a scientific research centre specializing in the investigation of the structures of solid matter and materials, as well as new materials and manufacturing techniques for photovoltaic cells. It has some 800 employees. A research reactor and an accelerator are avail-able. Of particular interest is the Residual Stress Centre. Details concerning the available instrumentation at HMI can be found in [107].

A.1.5 Other European Neutron Sources

Other available European sources of neutrons, can be found in Table A.1, which includes also the four previously mentioned sources

A.2 Synchrotron radiation sources

A.2.1 ESRF Grenoble (France)

The European Synchrotron Radiation Facility (ESRF) Grenoble (F) is the most powerful synchrotron radiation source in Europe. It is a joint facility supported and shared by 18 countries. At the ESRF, physicists work side by side with chemists and materials scientists. Moreover biologists, medical doctors, meteorologists, geophysicists and archaeologists have become regular users. Industrial applications are also growing, notably in the fields of pharma-ceuticals, cosmetics, petrochemicals and microelectronics. Among the largest and most powerful synchrotrons in the world, ESRF is the second after APS (USA) and before Spring-8 (Japan). The electrons reach an energy level of 6 GeV. Details on the available instrumentation at ESRF can be found in [109].

Table A.1. List of the European neutron sources, labelled by the name of the source, or the corresponding centre or laboratory (from [108])

Budapest Neutron Centre, AEKI, Budapest, Hungary
Berlin Neutron Scattering Center, Hahn-Meitner-Institut, Berlin, Germany
Center for Fundamental and Applied Neutron Research (CFANR), Rez-Prague, Czech Republic
Frank Laboratory of Neutron Physics, Joint Institute of Nuclear Research, Dubna, Russia
FRJ-2 Reactor, Forschungzentrum Jülich, Germany
FRM-II Research Reactor, Garching, Germany
GKSS Research Center, Geesthacht, Germany
Institut Laue Langevin, Grenoble, France
Interfacultair Reactor Instituut, Delft University of Technology, Netherlands
ISIS Pulsed Neutron and Muon Facility, Rutherford-Appleton Laboratory, Oxfordshire, UK
JEEP-II Reactor, IFE, Kjeller, Norway
Laboratoire Lon Brillouin, Saclay, France
Ljubljana TRIGA MARK II Research Reactor, J. Stefan Institute, Slovenia
Riso National Laboratory, Denmark
St. Petersburg Nuclear Physics Institute, Gatchina, Russia
Studsvik Neutron Research Laboratory (NFL), Studsvik, Sweden
Swiss Spallation Neutron Source (SINQ), Villigen Switzerland
RPI, Portuguese Research Reactor, Savacem, Portugal

A.2.2 ELETTRA Trieste (Italy)

ELETTRA is a multidisciplinary Synchrotron Light Laboratory in AREA Science Park, open to researchers operating in different basic and applied fields. The laboratory is equipped with ultra-bright light sources in the spectral range from UV to X-rays and offers a stimulating and competitive environment to researchers from all over the world. Details on the instrumentation available at ELETTRA can be found in [110].

A.2.3 HASYLAB Hamburg (Germany)

HASYLAB (Hamburger Synchrotronstrahlungslabor), with two synchrotrons and one free-electron laser based light source, is part of the national research institution DESY. HASYLAB researchers from universities, industries, and government laboratories investigate the fields of physics, biology, chemistry and crystallography, materials and geological sciences as well as medical applications. The European Molecular Biology Laboratory (EMBL) has established its own outstation on the DESY site for elucidating the structure of biomolecules. Details on the available instrumentation at HASYLAB can be found in [111].

A.2.4 SOLEIL Paris (France)

Soleil is a major French scientific facility which makes it possible to explore the microscopic structure of materials or living matter and to study their physical, mechanical and/or biological properties. The characteristics of this multidisciplinary tool have been defined in order to carry out a very wide range of activities in the fields of fundamental research as well as applied or industrial research. Details on the instrumentation available at SOLEIL can be found in [112].

A.2.5 DIAMOND Oxford (United Kingdom)

Diamond Light Source is a new scientific facility currently being built in South Oxfordshire on the Harwell Chilton science campus. It is housed in a futuristic doughnut-shaped building which covers the area of 5 football pitches. Diamond will ultimately host up to 40 cutting edge research stations, supporting the life, physical and environmental sciences. Diamond is a third generation 3 GeV (giga electron volt) synchrotron light source. Details on the instrumentation available at DIAMOND can be found in [113].

A.2.6 Other European synchrotron sources

A list of the other European synchrotron sources is reported in Table A.2, which includes also the five above mentioned sources

Table A.2. X-ray synchrotron sources in Europe [111]

Denmark:	ISA (Aarhus)
France:	ESRF (Grenoble), LURE (Orsay), Soleil (Orsay)
Germany:	ANKA (Karlsruhe), BESSY (Berlin), DELTA (Dortmund), ELSA (Bonn), HASYLAB (Hamburg)
Italy:	Elettra (Trieste), DAFNE (Frascati)
Spain:	ALBA (Barcelona)
Sweden:	MAX (Lund)
Switzerland:	SLS (PSI) (Villigen)
United Kingdom:	Diamond (Didcot), SRS (Daresbury)

References

1. ABAQUS (1998) Theory Manual, ver 5.8 edn Habbit, Karlsson, Sorensen, Inc
2. Aboudi J, Arnold S, Pindera MJ (1994) Response of a functionally graded composites to thermal gradients, Compos Eng 4, 1–18
3. Aboudi J, Pindera MJ, Arnold SM (1995) Thermo–inelastic response of functionally graded composites, Int J Solids Structures 32, 1675–1710
4. Aboudi J, Pindera MJ, Arnold SM (1996) Thermoelastic theory for the response of materials functionally graded in two directions, Int J Solids Structures 33, 931–966
5. Aboudi J, Pindera MJ, Arnold SM (1997) Microstructural optimization of functionally graded composites subjected to a thermal gradient via the higher-order theory, Composites Part B 28, 93–108
6. Abu Al-Rub RK, Voyiadjis GZ (2003) On the coupling of anisotropic damage and plasticity models for ductile materials, Int J Solids Struct 40, 2611–2643
7. Abu Al-Rub RK, Voyiadjis GZ (2006) A finite strain plastic-damage model for high velocity impact using combined viscosity and gradient localization militers: Part I – Theoretical formulation, Int J Damage Mech 15, 293–334
8. Akai T, Ootao Y, Tanigawa Y. (2005) Piezothermoelastic analysis of functionally graded piezoelectric cylindrical panel due to nonuniform heat supply in the circumferential direction, Proc Thermal Stresses'05, 709–712
9. Albertini G, Bruno G, Dunn BD, Fiori F Reimers W, Wright JS (1997) Mater Sci Eng A224:157–65
10. Albertini G, Ceretti M, Caglioti G, Fiori F, Monzani R, Viviani L (2001) Residual stresses in AA6082 shrink-lift systems: finite element calculations and neutron diffraction measurements, J Neutron Res 9, 459–467
11. Allen AJ, Hutchings MT, Windsor CG, Andreani C (1985) Adv Phys 34, 445
12. Ashbridge DA, Thorne MS, Rivers ML, Muccino JC, O'Day PA (2003) Image optimization and analysis of synchrotron X-ray computed microtomography (CμT) data, Comput Geosci 29, 823–836
13. Bansal Y, Pindera MJ (2003) Efficient formulation of the thermoelastic higher–order theory for functionally graded materials, J Thermal Stresses 26, 1055–1092
14. Bagri A, Eslami MR, Samsam-Shariat BA (2005) Coupled thermoelasticity of functionally graded layer, Proc Thermal Stresses'05, 721–724

15. Basista M (2001) Micromechanical and lattice modelling of brittle damage, Polish Acad Science, IFTR, Warsaw
16. Batterman BW, Cole H (1964) Dynamical diffraction of X rays by perfect crystals, Rev Mod Phys 36, 681–717
17. Bažant ZP (1984) Imbricate continuum and its variational derivation, J Eng Mech ASCE 110, 1593–1712
18. Bažant ZP (1994) Nonlocal damage theory based on micromechanics of crack interaction, J Eng Mech ASCE 120, 593–617
19. Bažant ZP, Pijaudier-Cabot G (1988) Nonlocal continuum damage, localization, instability and convergence, J Appl Mech ASME 55, 287–293
20. Benallal A (1989) Thermoviscoplasticité et endommagement des structures, Doctorat d'Etat, Universite Pierre et Marie Curie, Paris
21. Bentz DP, Martys NS, Stutzman PE, Levenson MS, Garboczi EJ, Dunsmuir J, et al (1995) X-ray microtomography of an ASTM C109 mortar exposed to sulfate attack, Mater Res Soc Symp Proc 370, 77–82
22. Bielski J, Skrzypek J, Kuna-Ciskał H (2006) Implementation of a model of coupled elastic–plastic unilateral damage material to finite element code, Int J Damage Mech 15, 1, 5–39
23. Björck A, Dahlquist G (1974) Numerical Methods, Englewood Cliffs, NJ, Prentice-Hall
24. Bouchard PJ, Fiori F, Treimer W (2002) Appl Phys A74, 1689–1691
25. Bruno G, Fiori F, Girardin E, Giuliani A, Koszegi L, Levy-Tubiana R, Manescu A, Rustichelli F (2001) Residual stress determination in several MMC samples submitted to different operating conditions, J Neutron Res 9, 107–117
26. Brückel T (2003) Elastic scattering from many-body systems, Lab Course Neutron Scattering, Forschungszentrum, Jülich
27. Buffière JY, Proudhon H, Ferrie E, Ludwig W, Maire E, Cloetens P (2005) Three dimensional imaging of damage in structural materials using high resolution micro-tomography, Nucl Instrum Methods Phys Res B 238, 75–82
28. Butler LG, Cartledge FK, Owens JW, Dowd B (1999) Synchrotron X-ray microtomography and solid-state NMR of environmental wastes in cement, in: Bonse U, editor, Developments in X-Ray tomography II, 97–104
29. Calbucci V, Fiori F, Manescu A, Rustichelli F, Stoebener K, to be published
30. Cantor CR, Schimmel PR (1980) Biophysical Chemistry, V.2 Techniques for the Study of Biological Structure and Function, Freeman, San Francisco
31. Caron I, Fiori F, Mesmacque G, Pirling T, Su M (2004) Expanded hole method for arresting crack propagation: residual stress determination using neutron diffraction, Physica B 350, 503–505
32. Cegielski M (2007) Numerical modelling of piston bottom in combustion engine made of metal-ceramic composite, MSc Thesis, Cracow Univ Technol, supervisor A. Ganczarski
33. Ceretti M, Coppola R, Lodini A, Perrin M, Rustichelli F (1995) High-resolution neutron diffractomer for internal stress measurements. Physica B 213–214:803–805
34. Chaboche J-L (1992) Damage induced anisotropy: On the difficulties associated with the active/passive unilateral condition, Int J Damage Mech 1, 148–171
35. Chaboche J-L (1993) Development of continuum damage mechanics for elastic solids sustaining anisotropic and unilateral damage, Int J Damage Mech 2, 311–332

36. Chaboche J-L (1997) Thermodynamic formulation of constitutive equations and application to the viscoplasticity of metals and polymers, Int J Solids Struct 34(18), 2239–2254

37. Chaboche J-L (1999) Thermodynamically founded CDM models for creep and other conditions, in Creep and Damage in Materials and Structures, Springer–Verlag, Wien New York, (Eds) H Altenbach, J Skrzypek, 209–283

38. Chaboche J-L, Rousselier G (1983) On the plastic and viscoplastic constitutive equations – P.1: Rules developed with internal variable concept, P.2: Application of internal variable concepts to the 316 stainless steel, J Pressure Vessel Technol 105, 153–164

39. Chaboche J-L (1977) Sur l'utilisation des variables d'etat interne pour la description de la viscoplasticité cyclique avec endommagemant, in Problems Non Linéaires de Méchanique 1977, Symposium Franco–Polonaise se Rhéologie et Méchanique, Cracovie, 137–159

40. Chaboche J-L, Lesne PM, Moire JF (1995) Continuum damage mechanics, anisotropy and damage deactivation for brittle materials like concrete and ceramic composites, Int J Damage Mech 4, 5–22

41. Chen XF, Chow CL (1995) On damage strain energy release rate Y, Int J Damage Mech 4, 251–236

42. Chen WF, Han DJ (1995) Plasticity for Structural Engineers. Springer-Verlag, Berlin Heidelberg

43. Chen B, Tong L (2004) Sensitivity analysis of heat conduction for functionally graded materials, Mater Design 25, 663–672

44. Chen B, Tong L (2004) Thermomechanically coupled sensitivity analysis and design optimization of functionally graded materials, Comput Methods Appl Mech Eng (in press)

45. Cho JR, Ha DY (2002) Volume fraction optimization for minimizing thermal stress in $Ni–Al_2O_3$ functionally graded materials, Mater Sci Eng A 334, 147–155

46. Cho JR, Shin SW (2004) Material composition optimization for heat-resisting FGMs by artificial neural network, Composites A 35, 585–594

47. Chow CL, Wang J (1987) An anisotropic theory of continuum damage mechanics for ductile materials, Eng Fract Mech 27(5), 547–558

48. Claesson T (2001) A medical imaging demonstrator of computed tomography and bone mineral densitometry, Stockholm, Universitetsservice US AB

49. Comi C (2001) A nonlocal model with tension and compression damage mechanisms, Eur J Mech A/Solids 20, 1–22

50. Comi C, Perego U (2004) Criteria for mesh refinement in nonlocal damage finite element analysis, Eur J Mech A/Solids 23, 615–632

51. Cordebois JP, Sidoroff F (1982) Anisotropic damage in elasticity and plasticity, J Mech Theor Appl, Num Special, 45–60, Coll Euromech 115, 1979 Grenoble

52. Cowley JM (1990) Diffraction Physics, North-Holland, Amsterdam

53. Dantz D, Genzel Ch, Reimers W ( ) Residual stress analysis in microwave sintered functionally gradient materials, Proceedings of the 4th Int Conf of FGM

54. Davidenkov NN, Spiridonova NE (1945) Analysis of stress state in the neck of tension specimen, Zavod Lab, XI, 6 (in Russian)

55. Dianoux AJ, Lander G (2003) Neutron Data Booklet, OCP Science, Grenoble, France

56. Di Prisco M, Ferrara L, Meftah F, Pamin J, de Borst R, Mazars J, Reynouard JM (2000) Mixed mode fracture in plain and reinforced concrete: some results on benchmark tests, Int J Fracture 103, 127–148

57. Domanus JC (1992) Practical Neutron Radiography, Kluwer Academic Publishers, Dordrecht, The Netherlands

58. Dornowski W, Perzyna P (2000) Localization phenomena in thermoviscoplastic flow processes under cyclic dynamic loading, Comput Assisted Mech Eng Sci 7, 117–160

59. Du S, Zagrodzki P, Barber JR, Hulbert GM (1997) Finite element analysis of frictionally excited thermoelastic instability, J Thermal Stresses 20, 185–201

60. Du et al. (2005) Mater Char 55, 75–82

61. Dufrénoy P, Weichert D (2001) A thermomechanical model for the analysis of disk brakes fracture, Proc Thermal Stresses'01, 463–466

62. Ebner R (2007) KMM-NoE NRT 3-5 Stress related phenomena (status report V8-1-2007)

63. Edwards L, Johnson MW, Priesmeyer HG, Rustichelli F, Withers PJ, Wright JS (1993) Precise measurements of internal stresses within materials using neutron residual stresses. In: Hauk V, et al. (eds) Proc 3rd Europ Conf Residual Stresses, Oberursel DGM

64. Egner H, Skrzypek JJ, Egner W (2006) Effect of characteristic length on non-local prediction of damage and fracture in concrete, J Theor Appl Mech 44, 485–503

65. Egner H, Juchno M, Kula M, Skrzypek J (2007) Numerical analysis of FG and TBC systems based on thermo-elasto-plastic-damage model, J Thermal Stresses, 30, 9, 977–1001

66. Elperin T, Rudin G (2002) Thermal stresses in functionally graded materials caused by thermal shock, Heat Mass Transfer 38, 625–630

67. Eshbach OW, Souders M (1975) Handbook of Engineering Fundamentals, John Wiley & Sons

68. Eshelby JD (1957) Proc Roy Soc, London A24, 376

69. Eslami MR, Bagri A, Samsam-Shariat BA (2005) Coupled thermoelasticity of functionally graded layer, Proc Thermal Stresses'05, 721–724

70. Eslami MR, Bahtui A (2005) Coupled thermoelasticity of thin cylindrical shells made of functionally graded materials, Proc Thermal Stresses'05, 729–732

71. Fan X, Lippmann H (1996) Elastic-plastic buckling of plates under residual stress, Proc Asia-Pacific Symp Adv Eng Plast Appl AEPA'96, Pergamon, Tokyo

72. Fanella D, Krajcinovic D (1988) A micromechanical model for concrete in compression, Eng Fracture Mech 29, 44–66

73. Fedehofer K, Egger H (1943) Knickung der auf Scherung beansprunchten Kreissingplatte mit veranderlicher Dicke, Ingr Arch 155–166

74. Feigin LA, Svergun DI (1987) Structure Analysis by Small-angle X-ray, Neutron Scattering, New York, Plenum Press

75. Finot M, Suresh S, Bull C, Sampath S (1996) Curvature changes during thermal cycling of a compositionally graded Ni–Al$_2$O$_3$ multi–layered material, Mater Sci Eng A205, 59–71

76. Fitzpatrick ME, Hutchings MT, Withers PJ (1997) Acta Mater 45, 12, 4867

77. Fujimoto T, Noda N (2000) Crack propagation in functionally graded plate under thermal shock, Arch Appl Mech 70, 377–386

78. Gan Z, Ng HW (2004) Experiments and inelastic finite element analyses of plasma sprayed graded coating under cyclic thermal shock, Mater Sci Eng A385, 314–324

79. Ganczarski A (2001) Problems of acquired anisotropy and coupled thermo-mechanical fields of CDM, Zeszyty Naukowe PK

80. Ganczarski A (2001) Computer simulation of transient thermo-damage fields in 3D structures, Proc ECCM-2001

81. Foryś P, Ganczarski A (2002) Modelling of microcrack closure effect, Proc Int Symp ABDM, Kraków-Przegorzały, Poland

82. Ganczarski A, Skrzypek J, Foryś P (2003) Anisotropic thermo–creep–damage in thick plate subjected to thermomechanical loading cycle, in: Onate, Owens (eds), Proc VII Int Conf on Comp Plasticity COMPLAS VII, CD

83. Ganczarski A, Barwacz L (2007) Low cycle fatigue based on unilateral damage evolution, Int J Damage Mech 16, 2, 159–177

84. Ganczarski A, Cegielski M (2007) Effect of continuous damage deactivation, Proc IV Int Symp Damage Mech Mater and Struct, Augustów, Poland

85. Garion C, Skoczeń B (2003) Combined model of strain-induced phase transformation and orthotropic damage in ductile materials at cryogenic temperatures, Int J Damage Mech 12, 4, 331–356

86. Gasik M, Lilius K (1994) Evolution of properties of W–Cu functional gradient materials by micromechanical model, Comput Mater Sci 3, 41–49

87. Genzel Ch, Stock C, Reiners W (2004) Mater Sci Eng A372:28

88. Glatter O, Kratky O (1982) Small-Angle X-ray Scattering, Academic Press, London

89. Glema A, Łodygowski T, Perzyna P (2000) Interaction of deformation waves and localization phenomena in inelastic solids, Comput Math Appl Mech Eng 183, 123–140

90. Guinier A, Fournet G (1955) Small Angle Scattering of X-ray, Wiley, New York

91. Guyot P, Cottignies L (1996) Acta Mater 44, 4161–4167

92. Haeffner DR, Almer JD, Lienert U (2005) The use of high energy X-rays from the Advanced Photon Source to study stresses in materials, Mater Sci Eng A399, 1-2, 120–127

93. Halm D, Dragon A (1996) A model of anisotropic damage by mesocrack growth; unilateral effect, Int J Damage Mech 5, 384–402

94. Hansen NR, Schreyer HL (1994) A thermodynamically consistent framework for theories of elastoplasticity coupled with damage, Int J Solids Struct 31, 359–389

95. Hansen NR, Schreyer HL (1995) Damage deactivation, J Appl Mech 62, 450–458

96. Hashimoto et al. (2005) Appl Surf Sci 241, 227–230

97. Hauk V (1997) Structural and Residual Stress Analysis by Nondestructive Methods. Elsevier

98. Hayakawa K, Murakami S (1997) Thermodynamical modeling of elastic–plastic damage and experimental validation of damage potential, Int J Damage Mech 6, 333–362

99. Hayakawa K, Murakami S (1998) Space of damage conjugate forces and damage potential of elastic–plastic–damage materials, Damage Mechanics in Engineering Materials (Eds) GZ Voyiadjis, J-W Ju, J-L Chaboche, Elsevier Science, Amsterdam, 27–44

100. Hearakovich CT, Aboudi J (1999) Thermal Effects in Composites, Thermal Stresses V, RB Hetnarski (Ed) Lastran Corporation Publ. Division, 1–142

101. Hernik S (2006) Stability analysis of MMC brake disk under thermomechanical loadings, MSc Thesis, Cracow Univ Tech, supervisor A. Ganczarski

102. Hernik S, Ganczarski A (2007) Notes on averaging of slowly-varying periodic functionally graded heat conductors, Proc Thermal Stresses07, Taiwan, Taipei, 287–290

103. Hones P, Martin N, Reguła M, Levy F (2003) Structural and mechanical properties of chromium nitride, molybdenum nitride, tungsten nitride thin films, J Phy D: Appl Phys 36, 1023–1029

104. http://www.ill.fr

105. http://www.isis.rl.ac.uk

106. http://www-llb.cea.fr

107. http://www.hmi.de/

108. http://pathfinder.neutron-eu.net

109. http://www.esrf.fr

110. http://www.elettra.trieste.it

111. http://www-hasylab.desy.de/

112. http://www.synchrotron-soleil.fr

113. http://www.diamond.ac.uk

114. http://www.hmi.de/bensc/instrumentation/instrumente/v12a/v12a_en.htm

115. Hutchings MT, Krawitz AD (1992) Advanced Research Workshop on Measurement of Residual and Applied Stress Using Neutron Diffraction. Proc NATO, Oxford, 18–22

116. IOP Publishing Limited (2004) High speed infrared cameras may boost the stopping power of express trains traveling at speeds of up to 300 km/h, http://optics.org/articles/news/10/9/11/1

117. Jackson JD (1999) Classical Electrodynamics (3rd edition), John Wiley & Sons Inc

118. Jacrot B (1976) The study of biological structures by neutron scattering from solution, Reg Prog Phys 39, 911

119. Jin Z-H, Paulino GH (2001) Transient thermal stresses analysis of an edge crack in a functionally graded material, Int J Fracture 107, 73–98

120. Jirasek M (1998) Nonlocal models for damage and fracture: comparison of approaches, Int J Solids Struct 35, 4133–4145

121. Ju JW (1989) On energy–based coupled elastoplastic damage theories constitutive modeling and computational aspects, Int J Solids Struct 25 (7), 803–833

122. Kak AC, Slaney M (1988) Principles of Computerized Tomographic Imaging, IEEE Press

123. Kapturkiewicz W (2006) Computer modelling of phase changes in alloys with particular attention to ADI casting, works IPPT PAN, 12–16

124. Kardjilov N, Hilger A, Manke I, Strobl M, Treimer W, Banhart J (2005) Industrial applications at the new cold neutron radiography and tomography facility of the HMI, Nuclear Instruments and Methods in Physics Research Section A: Accelerators, Spectrometers, Detectors and Associated Equipment, Volume 542, Issues 1-3, Proceedings of the Fifth International Topical Meeting on Neutron Radiography, 21 April 2005, 16–21

125. Kim JH, Paulino GH (2002) Mixed mode fracture of orthotropic functionally graded materials using finite elements and the modified crack closure method, Eng Fracture Mech 69, 1557–1586

126. Kim JH, Paulino GH (2002) Isoparametric graded finite elements for non–homogeneous isotropic and orthotropic materials, ASME J Appl Mech 69, 502–514

127. King A, Steuwer A, Woodward C, Withers PJ (2006) Effects of fatigue and fretting on residual stresses introduced by laser shock peening, Mater Sci Eng A435–436:12–18

128. Kobayashi H (2001) Basics of image plate for radiograms, Nondestr Test Eval 16, 245–54

129. Kokini K, Takeuchi YR, Choules BD (1996) Surface thermal cracking of thermal barrier coating owing to stress relaxation: zirconia vs. mullite, Surface and Coatings Technol 82, 77–82

130. Komlev V, Peyrin F, Mastrogiacomo M, Cedola A, Papadimitropoulos A, Rustichelli F, Cancedda R (2006) Kinetics of in-vivo bone deposition by marrow stromal cells into porous calcium phosphate scaffolds: a X-ray computed microtomography study, Tissue Eng 12, 12

131. Kompatscher M, Demé B, Kostorz G, Somsen C, Wassermann EFM (2002) Acta Mater 50, 1581–1586

132. Kostorz G (1982) Neutron Scattering, in Treatise on Materials Science and Technology, Vol 15, G Kostorz (ed.), Academic Press, London

133. Krajcinovic D (1989) Damage mechanics, Mech Mater 8, 117–197

134. Krajcinovic D (1996) Damage Mechanics, Elsevier, Amsterdam

135. Krajcinovic D, Fonseka GU (1981) The continuous damage theory of brittle materials, Part I: General theory, Trans ASME, J Appl Mech 48 (4), 809–815

136. Kuna-Ciskał H (1999) Local CDM based approach to fracture of elastic–brittle structures, Technische Mechanik 19, 351–361

137. Kuna-Ciskał H, Skrzypek J (2004) CDM based modelling of damage and fracture mechanisms in concrete under tension and compression, Eng Fracture Mechanics 71, 681–698

138. Ladeveze P (1983) On an anisotropic damage theory, in Failure Criteria of Structural Media, CNRS Int Coll No 351, Villard-de-Lans, ed. Boehler, Balkema, Rotterdam

139. Lame O, Bellet D, Di Michiel M, Bouvard D (2003) In situ microtomography investigation of metal powder compacts during sintering, Nucl Instrum Methods Phys Res B 200, 287–294

140. Landau LD, Lifshitz EM (1965) Quantum Mechanics: Non-relativistic Theory, Pergamon Press, Wiley, London

141. Landis EN, Nagy EN, Keane DT (2003) Microstructure and fracture in three dimensions, Eng Fract Mech 70, 911–25

142. Landis EN, Petrell AL, Lu S, Nagy EN (2000) Examination of pore structure using three dimensional image analysis of microtomographic data, Concrete Sci Eng 2, 162–9

143. Larsson C, Odén M (2004) X-ray diffraction determination of residual stresses in functionally graded WC-Co composites, International Journal of Refractory Metals and Hard Materials, 22, 4-5, July-September 2004, 177–184

144. Lee WY, Stinton DP, Berndt CC, Erdogen F, Lee Y–D, Mutasin Z (1996) Concept of functionally graded materials for advanced thermal barrier coating applications, J Am Ceram Soc 79, 3003–3012

145. Lemaitre J, Chaboche J-L (1985) Méchanique des Matériaux Solides, Dunod Publ, Paris

222   References

146. Lemaitre J (1986) Local approach of fracture, Eng Fracture Mech 25, 5/6, 523–537
147. Lemaitre J (1992) A Course on Damage Mechanics, Springer, Berlin
148. Lemaitre J, Chaboche J-L (1990) Mechanics of Solid Materials, Cambridge University Press, Cambridge, UK
149. Lehmann E, Vontobel P, Tobler L, Berger L, Deschler-Erb E, Voûte A (2002) Non-destructive Studies of Archaeological Objects Using Neutron Techniques, NUM Progress Report, http://num.web.psi.ch/reports/2002/pdf/e-19.pdf
150. Litewka A (1985) Effective material constants for orthotropically damaged elastic solid, Arch Mech 6, 631–642
151. Litewka A (1991) Damage and fracture of metals in creep conditions, Thesis of Poznań University of Technology, 250
152. Lorentzen T (1993) J Neutron Res 1, 13
153. Lubarda VA, Krajcinovic D, Mastilovic S (1994) Damage model for brittle elastic solids with unequal tensile and compressive strength, Eng Fracture Mech 49 (5); 681–697
154. Lukáš P, Vrána M, Šaroun J, Ryukhtin V, Vleugels J, Anné G Van der Biest O, Gasik M (2005) Neutron diffraction studies of functionally graded alumina/zirconia ceramics, Mater Sci Forum 492–493, 201–206
155. Manescu A, Giuliani A, Fiori F, Baretzky B (2006) Residual stress analysis in reed pipe brass tongues of historic organs. Mater Sci Forum 524–525:969–974
156. Mazars J, Pijaudier-Cabot G (1989) Continuum damage theory–application to concrete, J Eng Mech ASCE 115, 345–365
157. Mazars J, Pijaudier-Cabot G (1996) From damage to fracture mechanics and conversely: a combined approach, Int J Solids Struct 33, 3327–3342
158. Michalak B, Woźniak Cz, Woźniak M (2007) Modelling and analysis of certain functionally graded heat conductors, Archive Applied Mechanics (in print)
159. Miyazaki S, Ohmi Y, Otsuka K, Suzuki Y (1982) J Phys (Coll IV), 43, 255
160. Mura T (1987) Micromechanics of Defects in Solids, Dordrecht, Martinus Nijhoff Publishers
161. Murakami S, Liu Y (1995) Mesh-dependence in local approach to creep fracture, Int J Damage Mech 4, 230–250
162. Murakami S, Kamiya K (1997) Constitutive and damage evolution equations of elastic-brittle materials based on irreversible thermodynamics, Int J Solids Struct 39, 473–486
163. Murakami S, Ohno N (1980) A continuum theory of creep and creep damage, in Creep in Structures, ed Ponter ARS and Hayhurst DR, Springer, Berlin, 422–444
164. Naghdabadi R, Hosseini Kordkheili SA (2005) A finite element formulation for analysis of functionally graded plates and shells, Arch Appl Mech 74, 375–386
165. Noda N, Ishihara M, Yamamoto N (2004) Two-crack propagation paths in a functionally graded material plate subjected to thermal loading, J Thermal Stresses 27, 457–469
166. Noda N, Jin ZH (1993) Thermal stress intensity factors for a crack in a strip of a functionally graded materials, Int J Solids Struct 31, 1039–1056
167. Nooru-Mahammed MB (1992) Mixed mode fracture of concrete: an experimental approach, PhD Thesis, Delft Univ of Technol, 151
168. Nooru-Mahammed MB, Schlangen E, van Mier JGM (1993) Experimental and numerical study on the behaviour of concrete subjected to biaxial tension and shear, Adv Cement Based Mater 1, 22–37

169. Noyan IC, Cohen JB (1987) Residual Stress Measurements by Diffraction and interpretation, Material Research Engineering. Springer-Verlag

170. Nuzzo S, Peyrin F, Cloetens P, Baruchel J, Boivin G (2002) Quantification of the degree of mineralization of bone in three dimension using synchrotron radiation microtomography, Med Phys 19, 2672–81

171. Oleksy M, Skrzypek J (2007) Modelowanie cienkiej warstwy CrN na podłożu W300 przy cyklicznym wymuszeniu termicznym, Proc IV Int Symp Damage Mech Mat and Struct, Augustów, Poland

172. Oliver J, Huespe AE, Pulido MGD, Chaves E (2002) From continuum mechanics to fracture mechanics: the strong discontinuity approach, 69, 113–136

173. Ootao Y, Tanigawa Y, Ishimaru O (2000) Optimization of material composition of functionaly graded plate for thermal stress relaxation using a genetic algorithm, J Thermal Stresses 23, 257–271

174. Ootao Y, Akai T, Tanigawa Y (2005) Piezothermoelastic analysis of functionally graded piezoelectric cylindrical panel due to nonuniform heat supply in the circumferential direction, Proc Thermal Stresses'05, 709–712

175. Odqvist FKG (1966) Mathematical Theory of Creep and Creep Rupture, Oxford, Clarendon Press

176. Ottosen NS, Ristinmaa M (2005) The Mechanics of Constitutive Modeling, Elsevier, Oxford UK

177. Panier S, Dufrénoy P, Weichert D (2001) Macroscopic hot-spots occurrence in fictional organs, Proc Thermal Stresses'01, 621–624

178. Paulus MJ, Gleason SS, Kennel SJ, Hunsicker PR, Johnson DK (2000) High resolution x-ray computed tomography: an emerging tool for small animal cancer research, Neoplasia 2, 62–70

179. PCGPACK User's Guide, New Haven: Scientific Computational Associates, Inc

180. Pijaudier-Cabot G, Bažant ZP (1987) Nonlocal damage theory, J Eng Mech ASCE 113, 1512–1533

181. Poterasu F, Sugano Y (1993) An improved solution to thermoelastic material design in functionally graded materials: Scheme to reduce thermal stresses, Comput Mech Appl Mech Eng 109, 377–389

182. Pranzas et al. (2006) Physica B 385–386, 630–632

183. Press WH, Teukolsky SA, Vetterling WT, Flannery BP (1992) Numerical Recipes in Fortran, Cambridge University Press

184. Priesmeyer HG, Larsen J, Meggers D (1994) J Neutron Res. 2:31

185. Proceedings of the 7th European Conference on Residual Stresses   Berlin, Germany 13-15 September 2006

186. Rabin BH, Williamson RL, Bruck HA, Wang X-L, Watkins TR, Feng Y-Z, Clark DR (1998) Residual strains in an Al2O3-Ni joint bonded with a composite interlayer: experimental measurements and FEM analysis, J Am Ceram Soc 81, 1541–1549

187. Rangaraj S, Kokini K (2003) Interface thermal fracture in functionally graded zirconia/mullite-bond coat alloy thermal barrier coatings, Acta Mater 51, 251–267

188. Rangaraj S, Kokini K (2003) Estimating the fracture resistance of functionally graded thermal barrier coatings from thermal shock tests, Surface Coatings Technol 173, 201–212

189. Rangaraj S, Kokini K (2004) A study of thermal fracture in functionally graded thermal barrier coatings using a cohesive zone model, J Eng Mater Technol 126, 103–115

190. Rychlewska J, Szymczyk J, Woźniak Cz (2004) On the modelling of the hyperbolic heat transfer problems in periodic lattice-type conductors, J Thermal Stresses 27, 825–843

191. Rymarz C (1993) Mechanika ośrodków ciągłych, PWN, Warszawa

192. Saanouni K (1988) Sur l'analyse de la fisuration des milieux élastoviscoplastiques par la théorie de l'endommagement continu, Doctorat d'Etat, Université de Compiegné

193. Saanouni K, Forster C, Ben Hatira F (1994) On the inelastic flow with damage, Int J Damage Mech 3(2), 140–169

194. Saburi T, Tatsumi T, Nenno S (1982) J Phys, (Coll IV), 43, 261

195. Salomé M, Peyrin F, Cloetens P, Odet C, Laval-Jeantet AM, Baruchel J, Spanne P (1999) A synchrotron radiation microtomography system for the analysis of trabecular bone samples, Med Phys 26, 2194

196. Saouridis C, Mazars J (1992) Prediction of failure and size effect in concrete via a bi-scale damage approach, Eng Comput J 9, 329–344

197. Schillinger B, Lehmann E, Vontobel P (2000) Physica B59, 276–278

198. Schneider U (1988) Concrete at high temperatures. A general review, Fire Safety J 13, 55–68

199. Schreiber J, Neubrand A, Wieder T, Shamsutdinov N (2000) Distribution of Macro- and Micro-Stresses in W-Cu FGM, Functionally Graded Materials, 603–610

200. Sears VF (1992) Neutron scattering length and cross sections, Neutron News, 3, 26-37

201. Shashiah R, Eslami MR (2003) Thermal buckling of functionally graded cylindrical shell, J Thermal Stresses 26, 277–294

202. Schulz U, Bach FW, Tegeder G (2003) Graded coating for thermal, wear and corrosion barriers, Mater Sci Eng A 362(1–2), 61–80

203. Simo JC, Ju JW (1987) Strain– and stress–based continuum damage models; I–Formulation, II–Computational aspects, Int J Solids Struct 23, 821–869

204. Skrzypek JJ (1999) Material damage models for creep failure analysis and design of structures, in Creep and Damage in Materials and Structures, H Altenbach, J Skrzypek (Eds) Springer Wien New York, 97–166

205. Skrzypek J (2006) Podstawy Mechaniki Uszkodzeń, Wyd Politechniki Krakowskiej

206. Skrzypek JJ, Ganczarski A (1998) Modeling of damage effect on heat transfer in time–dependent nonhomogeneous solids, J Thermal Stresses 21, 205–231

207. Skrzypek JJ, Kuna-Ciskał H (2003) Anisotropic elastic–brittle–damage and fracture model base on irreversible thermodynamics, in: Skrzypek, Ganczarski (eds), Anisotropic Behaviour of Damage Materials, pp.143–184, Springer, Berlin–Heidelberg–New York

208. Skrzypek JJ, Kuna-Ciskał H, Bielski J (2004) Damage acquired anisotropy in elastic–plastic materials, Proc 21 ICTAM, Warsaw, Paper SM4L 11796

209. Skrzypek JJ, Kuna-Ciskał H, Egner W (2005) Non-local description of anisotropic damage and fracture in elastic-brittle structures under plane stress conditions, Int Conf Fracture ICF11, Torino, 20–25, CD

210. Skrzypek J, Kuna–Ciskał H, Egner W (2005) Non–local description of damage and fracture in double notched concrete specimen under plane stress conditions, Proc III Symp Mechaniki Zniszczenia Materiałów i Konstrukcji, Augustów, 381–384

211. Sladek J, Sladek V, Zhang C (2003) Transient heat conduction analysis in functionally graded materials by the meshless local boundary integral equation method, Comput Mater Sci 28, 494–504

212. Sneddon IN, Lockett FJ (1960) On the steady-state thermoelastic problem for the half-space and the thick plate, Quart Appl Math 18, 2, 145–153

213. Spencer AJM (1971) Theory of invariants, in Continuum Physics, Ed EC Eringen, Acad Press, New York, 239–353

214. Squires GL (1978) Introduction to the Theory of Thermal Neutron Scattering, Cambridge University Press, Cambridge

215. Tamma KK, Avila AF (1999) An integrated micro/macro modeling and computational methodology for high temperature composites, Thermal Stresses V, RB Hetnarski (Ed), Lastran Corporation Publ Division, 143–256

216. Tanigawa Y (1993) Theoretical approach of optimum design for a plate of functionally gradient materials under thermal loadings, in Thermal Shock and Thermal Fatigue Behavior of Advanced Ceramics, Schneider GA, Petzow G eds., 171–180, Kluwer Acad Publ Dordrecht

217. Tanigawa Y (1995) Some basic thermoelastic problems for nonhomogeneous structural materials, Appl Mech Rev 46, 287–300

218. Tomota Y, Kuroki K, Mori T, Tamura I (1976) Tensile deformation of two–ductile–phase alloys: flow curves of $\alpha \rightarrow \gamma$ Fe–Cr–Ni alloys, Mater Sci Eng 24, 85–94

219. Ueda S (2001) Thermoelastic analysis of W–Cu functionally graded materials subjected to a thermal shock using a micromechanical model, J Thermal Stresses 24, 19–46

220. Ueda S (2001) Elastoplastic analysis of W–Cu functionally graded materials subjected to a thermal shock by micromechanical model, J Thermal Stresses 24, 631–649

221. Ueda S (2001) Thermal shock fracture in a W–Cu divertor plate with a functionally graded nonhomogeneous interface, J Thermal Stresses 24, 1021–1041

222. Ueda S (2002) Transient thermal singular stresses of multiple cracking in a W–Cu functionally graded plate, J Thermal Stresses 25, 83–95

223. Ueda S., Gasik M. (2000): Thermal–elasto–plastic analysis of W–Cu functionally graded materials subjected to a uniform heat flow by micromechanical model, J Thermal Stresses 23, 395–409

224. Velhinho A, Sequeira PD, Martins R, Vignoles G, Braz Fernandes F, Botas JD, Rocha LA (2003) X-ray tomographic imaging of Al/SiC_p functionally graded composites fabricated by centrifugal casting, Nuclear Instruments and Methods in Physics Research, Section B: Beam Interactions with Materials and Atoms, 200, January 2003, 295–302

225. Voyiadjis GZ, Deliktas B (2000) A coupled anisotropic damage model for inelastic response of composite materials, Comput Meth Appl Mech Eng 183, 159–199

226. Voyiadjis GZ, Park T (1997) Local and interfacial damage analysis of metal matrix components using finite element method, Eng Fracture Mech 56 (4), 483–511

227. Voyiadjis GZ, Park T (1999) Kinematics description of damage for finite strain plasticity, Int J Eng Sci 37, 803–830

228. Voyiadjis GZ, Deliktas B, Aifantis EC (2001) Multiscale analysis of multiple damage mechanics coupled with inelastic behavior of composite materials, J Eng Mech 127, 636–645

229. Voyiadjis GZ, Abu Al-Rub RK (2002) Length scales in gradient plasticity, Proc IUTAM Symp on Multiscale Modelling and Characterisation of Elastic–Inelastic Behavior of Engineering Materials, Ahzi S, Cherkanoui M (Eds), Marocco, Kluwer Acad Publ

230. Voyiadjis GZ, Abu Al-Rub RK (2006) A finite strain plastic-damage model for high velocity impact using combined viscosity and gradient localization mimiters: Part II – Numerical aspects and simulations, Int J Damage Mech 15, 335–374

231. Wang WM, Sluys LJ, de Borst R (1996) Interaction between material length scale and imperfection size for localization phenomena in viscoplastic media, Eur J Mechanics A/Solids 15, 447–464

232. Wang B-L, Han JC, Du SY (2000) Crack problems for functionally graded materials under transient thermal loading, J. Thermal Stresses 23, 143–168

233. Wang B-L, Mai Y-W, Zhang X-H (2004) Thermal shock resistance of functionally graded materials, Acta Mech 52, 4961–4972

234. Wawrzynek PA, Ingraffea AR (1991) Discrete modeling of crack propagation: theoretical aspects and implementation issues in two and three dimensions, Report 91–5, School of Eng and Environmental Univ Cornell Univ

235. Wells GN, Sluys LJ (2000) Application of embedded discontinuities for softening solids, Eng Fract Mech 65, 263–281

236. Wilding M, Sumner DY, Lesher CE, Shields KE, Reap D, Richards WJ (2002) Neutron computed tomography reconstructions of microbial structures in carbonate and silicate rocks, Presented at the Second Astrobiology Science Conference, April 7-11, NASA AMES Research Center, USA, http://www.astrobiology.com/asc2002/abstract.html?ascid=133

237. Withers PJ et al (1994) Proc Int Symp On Advanced Materials for Light Weight Structures, Noordwijk

238. Woźniak Cz, Wierzbicki E (2000) Averaging techniques in thermomechanics of composite solids. Tolerance averaging versus homogenization, Wydawnictwa Politechniki Częstochowskiej, Częstochowa

239. www.bohler.de

240. www.matweb.com

241. Yamanouchi M, Hirai T, Shiota I (1990) Overall view of the P/M fabrication of functionally gradient materials, Proc First Int Symp Functionally Garadient Materials, Eds Yamanouchi et al, Sendai, Japan, 59–64

242. Yi Y-B, Barber JR, Zagrodzki P (2000) Eigenvalue solution of thermoelastic instability problems using Fourier reduction, The Royal Society

243. Yildirim B, Erdogan F (2004) Edge crack problems in homogeneous and functionally graded material thermal barrier coatings under thermal loading, J Thermal Stresses 27, 311–329

244. Ylinen A (1956) A method of determining the buckling stress and required cross-sectional area for centrally loaded straight columns in elastic and inelastic range, Mémoires, Association Internationale des Ponts et Charpents, Zürich, 16, 529–550

245. Zhu YY, Cescotto S (1995) A fully coupled elasto–visco–plastic damage theory for anisotropic materials, Int J Solids Struct 32 (11), 1607–1641

Index

Printing: Krips bv, Meppel, The Netherlands
Binding: Stürtz, Würzburg, Germany